the Science of the Soul

SCIENTIFIC EVIDENCE OF HUMAN SOULS

KEVIN T. FAVERO

Beaver's Pond Press, Inc.
Edina, Minnesota

ISBN: 1-59298-055-4

Library of Congress Catalog Number: 2004110953

Book design and typesetting by Mori Studio

Printed in the United States of America

First Printing: August 2004

08 07 06 05 04   5 4 3 2 1

7104 Ohms Lane, Suite 216
Edina, MN 55439
(952) 829-8818
www.beaverspondpress.com

to order, visit:
www.ScienceofSouls.com or www.BookHouseFulfillment.com
or call 1-800-901-3480. Reseller discounts available.

To my wife, Gwendolyn, the love of my life, and to my children, Renee, Rachel, Reed, Rosemary, and Rebecca, for your loving support and insightful contributions.

To my parents, Renato Anthony Favero and Rita Marie Groesch Favero, for participating with God in giving me life and for teaching me God's love by your lives.

To my brothers, sisters, my wife's sister, and your spouses, Tom and Donna Favero Pforr, Paul and Phyllis Favero, Dan and Janice Favero Viele, Jim and Colette Favero Ellenberg, Daniel and Melinda Favero, Thomas and Carol Favero Burke, Jeremiah and Lynn Favero, Michael and Mary Favero Burke, Robert and Terri Long, and to my nephews, nieces, and your spouses for your loving support.

To my aunts and uncles, Robert Favero, Remus and Marie Favero, Rudy and Eileen Favero, Mary Favero, Irene Favero Bourgasser, Hilda Favero Micheletti, Lubert and Lorene Groesch, Betty Groesch, Paul and Loretta Groesch Mahoney, Dolores Groesch Owensby, Michael and Margaret Groesch Burns, and to my cousins and your spouses for your loving support.

In memory of the members of my family whose souls are no longer joined to this natural realm and who are in the care of God:

my grandparents, Fortunato and Pasqua Favero, Edward and Frances Groesch;

my sister, Marsha Ellen Favero;

my aunts and uncles, Raymondo Favero, Riserio and Josephine Favero, Massimigliano and Eda Favero Meneghetti, Joseph and Erma Favero Cohn, Joseph and Etalia Favero Plesh, Dolcedo Micheletti, Andrew Bourgasser, Joseph and Mary Groesch, Rev. Edward Groesch, Louis Groesch, Roy Bowman, and John Owensby;

my cousins, William Micheletti, Michael Burns, Gregory Groesch, and Gary Groesch, who dedicated his life to help the people of New Orleans, and my cousin's spouse, James Kelly.

Table of Contents

Acknowledgments

This book is dedicated to the memory of the following philosophers and scientists who developed some of the basic concepts on which this book is based:

René Descartes (1596–1650), philosopher and mathematician;

Sir John Eccles (1903–1997), brain researcher and 1963 winner of the Nobel Prize for physiology; and

Mortimer Adler (1902–2001), teacher and philosopher.

ACKNOWLEDGMENT OF HIGH SCHOOL TEACHERS AND ADMINISTRATORS

This book is also dedicated to my high school teachers and administrators at Divine Heart Seminary in Donaldson, Indiana who provided me with the inspiration and the tools to pursue this endeavor:

Charles Adam	Fr. James Alexander, S.C.J.
Anthony Bruder	Fr. Tom Cassidy, S.C.J.
Fr. Francis Clancy, S.C.J.	Fr. Joseph Coyle, S.C.J.
Fr. Bernard Galic, S.C.J.	Fr. Tom Garvey, S.C.J.
Br. Larry Gauthier, S.C.J.	George Grob
Fr. Justin Guiltnane, S.C.J.	Fr. Joseph Haselbauer, S.C.J.
John Hennessy	Fr. Robert Holko, S.C.J.
Robert Lenz	John Madden
Br. Peter Mankins, S.C.J.	John McEvilly
Peter Meade	Dale Mitchell
Frederick Moffet	Geoffrey Pratte
David Prestipino	Dn. Marion Quagliariello
Fr. Bernard Rosinski, S.C.J.	Donald Schmid

Br. Gerald Selenke, S.C.J.	Alphonse Signorino
Br. Francis Snider, S.C.J.	Fr. Thomas Sheehy, S.C.J.
Fr. James Steffes, S.C.J.	Br. Conrad Thelan, S.C.J.
Mark Turner	Fr. Michael van der Peet, S.C.J.
Br. Ray Vega, S.C.J.	James Wake
Fr. Dominic Wessel, S.C.J.	Fr. Frank Wittouck, S.C.J.

ACKNOWLEDGMENT OF REVIEWERS

Special thanks and acknowledgment is also due to the following professors and teachers who provided valuable review and corrections to this manuscript. I cannot express enough how helpful their insights and suggestions have been.

Dr. Jon K. Beane, Ph.D. in philosophy, University of Notre Dame, teacher at Trinity at River Ridge, Bloomington, Minnesota;

Dr. John R. Goodreau, Ph.D. in philosophy, The Catholic University of America, teacher at Trinity at River Ridge, Bloomington, Minnesota;

Dr. John Vogel, Ph.D. in mathematics, Courant Institute of Mathematics, New York University; B.E. civil engineering, The Cooper Union; teacher at Trinity at River Ridge, Bloomington, Minnesota;

Dr. William Monsma, Ph.D. in physics, University of Colorado, masters of divinity degree, Calvin Theological Seminary. Founder of The Maclaurin Institute; adjunct professor at the University of Minnesota;

Dr. Christopher J. Edge, Ph.D. in atomic physics/laser spectroscopy, University of Virginia, chief scientist, Kodak Polychrome Graphics, St. Paul, Minnesota;

Dr. Martha Phillips, Ph.D. in biology, University of Minnesota; professor of biology at the College of St. Catherine, St. Paul, Minnesota;

Dr. Cindy Norton, Ph.D. in biology, University of Iowa; professor and biology department chair at the College of St. Catherine, St. Paul, Minnesota; and

Dr. Richard Bellon, Ph.D. in history, University of Washington, with a specialty in the history of natural history and systematic biology; visiting assistant professor in history of science and technology, University of Minnesota.

Special thanks to Dr. Ray Bressan, Ph.D. in biology, Colorado State University; distinguished professor of molecular biology at Purdue University; my boyhood friend in Springfield, Illinois with whom I shared many hours peering through microscopes during our teenage years and with whom I shared the emotional devastation of the collapse of the Chicago Cubs in 1969.

The review of this manuscript or portions of this manuscript by the above teachers and professors does not imply that they agree with or endorse any specific idea or statement contained herein. It goes without saying that any errors are the responsibility of the author.

Thanks also to Cindy Rogers, my editor, Sid Korpi, my proofreader, Morrie Lundin, my typesetter, and Milt Beaver Adams of Beaver's Pond Press, Inc. for many valuable suggestions and comments.

Thanks to Jack Caravela and Jay Monroe of Mori Studio, Inc. for an inspired cover and book design.

Thanks to William L. Porter for giving me the opportunity to work as an engineer and consultant for R.W. Beck and the opportunity to hone my technical writing skills by preparing reports for clients in the utility and energy industries.

Thanks to the members of St. Richard's Catholic Community who willingly participated in taking the free will test, including Corey Brown, Pat de Lozier, Joyce Eppel, Rebecca Favero, Sister Lynore Girmscheid, Chris Norby, Mary Smith, and Ruben Soruco.

ARLO & JANIS reprinted by permission of United Feature Syndicate, Inc..

Preface or Gas in a Bottle

It began that day in high school when our English teacher asked what we thought about the human soul and whether it was like a "fart in a bottle."[1] That got quite a laugh from a group of high school boys who were more familiar with gas than with the nature of human souls. Nevertheless, I thought it was an interesting question and I vowed to study the matter as I got more time. I was taught that human souls are spiritual and supernatural. This meant that souls are immaterial, are not natural phenomena, and are not subject to the laws of nature. I was also taught that the soul includes the will and the intellect. However, other than this cursory description, the nature of human souls was not investigated in any detail. Studying for typical school courses both in high school and college did not leave much time to study the soul, and I felt I had more important things to do with my time after my class work had been prepared. Also, I eventually saw the idea of spiritual souls come under serious attack because they seemed to be contrary to the theories and scientific laws espoused in the science and electrical engineering courses I took at the University of Illinois.

I learned from those courses that all human thought is produced by electrochemical activity in the brain which operates as a very complex biological computer. My biology course taught that the human brain itself is the result of millions of years of evolution. The professor taught that the human brain is like the brain of other creatures only much more developed and complex and therefore capable of more sophisticated mental processes.

---

1. The teacher was John Madden who went on teach Latin and Greek at the University of Montana in Missoula from 1975 to 1998. More recently, he has been teaching philosophy at Cerritos College in Norwalk, California. Teachers, take note that the seed for this book was planted by a probing question in high school.

Faith was Challenged

I felt devastated that it appeared my long-held belief in God, supernatural souls, and eternal life was totally discredited by scientific principles. Although my faith was challenged, I took support from Thomas Aquinas and reasoned that "true" science and "true" religion would not be contradictory. I also believed that if there were anything valid in religion, my study of science would eventually lead me to those religious principles. This book describes the results of my scientific investigation into the existence and nature of supernatural souls and God. I am overjoyed to report that I have concluded science and faith are not contradictory and do lead to the same truths. More importantly, I have found:

- *there is overwhelming scientific evidence which I believe[2] leads to the logical conclusion that each human has a supernatural soul.[3]*

The evidence I rely upon is based on well-accepted mainstream scientific principles. It is not based on paranormal occurrences, parapsychology, or near-death experiences. Likewise, I have found that to not believe in supernatural souls is an unscientific position. This book and subsequent books will present a comprehensive world view based on the scientific evidence that all natural phenomena interact according to rules, the evidence that each human being can make willful choices, and the logical conclusion that each person has a supernatural soul. I have tried to identify the relevant trends in each applicable area of study and human endeavor that are affected by this evidence and the subsequent conclusion. In doing this, I have had to draw on nearly everything I have learned and every course I have studied in high school and college. I relate this to provide encouragement to students everywhere to take advantage of the opportunity to learn all that they can in every course they take.

---

2. As described in the chapter "Introduction and Basic Premise," each person must decide what conclusions are to be drawn from evidence and what he or she will believe. The beliefs I express in this book are based on the evidence presented and the conclusions I have drawn from such evidence.

3. And I have discovered that a spiritual soul is decidedly not like a fart in a bottle.

No Math

I have kept the use of math in this book to a bare minimum. An understanding of math is not needed to follow the basic theories presented in this book. A simple understanding of logic and common sense are all that are needed. To the extent I have included mathematical analyses, I have put them in appendices at the end of the respective chapters so that they would not unnecessarily encumber those readers that do not want to bother with the math. Even the chapter "Math" does not have mathematical analyses. The "Math" chapter deals only with mathematical concepts (described in everyday language) and covers a new mathematical theorem formulated in the 1930s that provides additional evidence for the existence of supernatural souls.

Physics to the Rescue

My analysis is based on well-established scientific principles, including the concepts of quantum physics developed in the early 1900s. How, you might ask, do quantum physics and mathematical concepts provide evidence for the existence of supernatural souls? Are humans the result of evolution? Does God exist? Our investigation into these and other relevant matters begins in chapter one.

Depression

By showing that there is sound scientific evidence which leads to the logical conclusion that humans have supernatural souls, I hope to free some people from their nihilistic belief that "this material world is all there is." I believe that a lack of belief in supernatural souls and in life after death likely causes depression in some people. As described by Viktor Frankl in his 1959 book *Man's Search for Meaning*, this type of depression is due to an "existential vacuum."[4] As anecdotal evidence, I can offer my

---

4. Frankl (1959) 1984 ed., pg. 128.

own feelings of depression when I was faced in college with the apparent scientific fact that human thought was only a complex collection of electrochemical reactions inside a human brain and that human souls do not exist. I find my life much more joyful now that I can again believe in supernatural souls based not only on faith but also on science.

In the 1997 book *Finding Flow: The Psychology of Engagement with Everyday Life*, author Mihaly Csikszentmihalyi describes the observations of a psychoanalyst at the beginning of the 1900s. The psychoanalyst discovered that he had an inordinate number of patients who suffered anxiety and depression on Sundays.[5] He attributed this to the fact that, at that time in history, people worked every day except Sunday and identified so closely with their work that they felt aimless and goalless on Sundays when they were not at work.

In the 1990 book *The Quantum Self*, physicist Danah Zohar describes how the alienation and depression that Bertrand Russell, a famous mathematician and philosopher of the 1900s, felt resulted from believing in only a material existence:

> *The world which science presents for our belief . . . [tells us] that man is the product of causes which had no pre-vision of the end they were achieving; that his origin, his growth, his hopes and fears, his loves and his beliefs, are but the outcome of accidental collocation of atoms; that no fire, no heroism, no intensity of thought and feeling, can preserve the individual life beyond the grave; that all the labours of the ages, all the devotion, all the inspiration, all the noonday brightness of human genius, are destined to extinction in the vast death of the solar system, and that the whole temple of man's achievement must inevitably be buried beneath the debris of a universe in ruins . . .*[6]

I think that Professor Russell presented a misleading description of what the world of science teaches. Science only seeks to

---

5. Csikszentmihalyi (1997), pg. 65.

6. Zohar (1990), pg. 19 as quoted from Russell, Bertrand, "A Free Man's Worship," *In Mysticism and Logic.* New York: Doubleday Anchor, 1957, pg. 45.

discover the laws that govern the natural world. Science makes no claim about the existence or nature of the supernatural realm. I would agree with Professor Russell's gloomy assessment only if I were to also believe that humans are a mere "collocation of atoms" governed by the laws of physics and doomed to ultimate extinction with the end of the solar system. Although it would be difficult to prove, I believe a significant amount of sadness and depression is caused by a lack of an eternal goal and a fear that this material world is all there is. Some of this fear might be happening at an unconscious level and the person affected might not even be aware of it. If this material world is all there is, then any intermediate goal pursued or achieved during our meager existence between birth and death becomes meaningless to some people. To them, life takes on the image of rearranging the deck chairs on the Titanic: something nice to do for the time being but without any long-term importance or meaning.

THE SPIRITUAL MILLENNIUM

In this book, I intend to provide scientific evidence I believe leads to the conclusion that human beings are more than just material bodies, more than just a "collocation of atoms." By providing evidence that leads to the conclusion that humans have supernatural souls, I hope to give hope and joy to those who currently only look at this life as a material existence. Based on this evidence, I hope people will be able to come to the belief that we are essentially spiritual rather than material beings. Such belief could dispel the anxiety some people feel about death and about whether life has meaning. Such belief could also give people a goal that transcends this world. This ultimate goal would be that of having ourselves changed spiritually so that we could live more fulfilled lives and be ready to enjoy what awaits us on the other side of death. With such a goal, we could be assured of having plenty of useful things to do during all periods of our lives:

- There will always be areas of our own spirituality that need improving.

- There will always be other people at various stages of spiritual development whom we can help and/or look to for guidance.
- There will always be other people who need material help. By helping them, we will fulfill the spiritual commandment to love our neighbor.

With this refocus, I am hopeful people can avoid the anxiety and depression that comes from lacking a goal. They can redirect their efforts to making sure all the decisions in their lives have a spiritual basis. I hope this millennium will be a spiritual millennium with more and more people coming to believe humans are spiritual beings with supernatural souls.

I do not mean to imply that this refocus will cure all depression in the world. There are undoubtedly as many reasons for depression as there are people in the world, including biochemical causes. I certainly do not intend to imply that the suffering many people endure due to depression is insignificant or easily cured. However, I nonetheless believe that there is a significant component of depression that is due to an underlying anxiety and fear that life ends with death. This is the depression I hope to relieve, at least in part, by providing evidence that leads to the conclusion that humans do have supernatural souls.

It's Good to Be Here!

With this renewed hope in eternal life and joy for living I hope people will feel that It's Good to Be Here!, that it's good to be alive. It is my belief that with life comes the opportunity to choose God, to choose love, to choose to know God's will for us, and to choose eternal life.

Guilt

Another source of anxiety and depression is guilt over our sins and moral failures. I believe that we become guilty by using our free will to reject God and to violate His commandment to love

Him and our neighbors. I believe that free will is a supernatural phenomenon and that we need a spiritual source of healing and forgiveness for our sins. By providing evidence that the supernatural world exists, I hope to make people aware of the healing power available to relieve them of the burden of guilt.

IS THERE A GOD?

Once the evidence is presented that leads to the conclusion that human supernatural souls exist, the next logical conclusion is that these souls cannot suddenly appear out of nowhere. If supernatural souls exist, it is logical to believe that they would need to be created by a loving, personal God who has performed and continues to perform a monumental task of creating these billions of souls and joining each of them to a human body. It would follow then that a God of love would create us with the intent that we ultimately enjoy eternal happiness with Him after our short period of trial on this earth. The next logical step is that we have to start thinking about what we need to do to attain eternal life and how we should live our lives to please our Maker.

SUNDAY ANXIETY

It seems ironic to me that the psychoanalyst referenced earlier would discover people having anxiety due to free time on Sunday. Resting on the seventh day was originally prescribed in the book of Exodus in the Bible to give people a chance to rest from their daily burdens and refocus their thinking on God and spiritual matters. Of course, if you have discarded the need for religion and spirituality and have based your worth on your job, then free time on Sunday does not provide a time for spiritual rest and refocus. Rather, it becomes a burden by taking you away from your job that provides meaning in your life and the way of achieving your material goals. The free time also gives you time to think how having only material goals is not ultimately very satisfying.

The Mind-Body Connection

There is growing evidence that the health of the human mind and body are intricately interconnected. Health is dependent not only on what is eaten by the body but also on what is thought by the mind. The existence of supernatural souls in no way contradicts this understanding. The soul-body connection can be as intricate as any envisioned for a mind-body connection. The health of the body and soul are likewise intertwined. I will focus on only one aspect of the soul (free will), but even this one aspect demonstrates the complex interrelationship between the soul and the body. During every moment of our lives, we continually balance between the things that our bodies do automatically and the free choices that we interject in infinite gradations.

A Science-Based Faith

By providing scientific evidence I believe leads to the conclusion that supernatural souls exist, I hope to give people the ability to base belief in a supernatural world on scientific principles as well as religious "faith." This would provide each of us with the ability to make an intellectual commitment to spirituality. Hopefully, this will help relieve any lingering doubts that might creep into our lives and the associated anxiety that religion is just the "opiate of the masses" as Karl Marx would have us believe. I invite you to begin your journey to investigate the scientific basis for believing in supernatural souls in Chapter 1.

Out of the night that covers me,
Black as the pit from pole to pole,
I thank whatever gods may be
For my unconquerable soul.

In the fell clutch of circumstance
I have not winced nor cried aloud.
Under the bludgeonings of chance
My head is bloody, but unbowed.

Beyond this place of wrath and tears
Looms but the horror of the shade.
And yet the menace of the years
Finds and shall find me unafraid.

It matters not how straight the gate
How charged with punishments the scroll,
I am the master of my fate;
I am the captain of my soul.

—*"Invictus" by William Henley*

Chapter One
INTRODUCTION AND BASIC PREMISE

The most fundamental questions we face as human beings are: what is the nature of a human being, why are we here, and what happens after our bodies die? Since the earliest times of recorded history, some philosophers and religious people have taught that each human being has a spirit or a soul that is different and distinct from the physical body. The soul was considered by some to be the essence of the person, the part that makes the person who he or she is: the personality, the memory, and the will. It was also considered to be the part of the person that remains even after the death of the body. Many religions, in addition to teaching about the souls of human beings, also taught about other spirits including a Supreme Spirit (who is called by many names).

Some religions had many gods and stories that explained occurrences in nature in terms of divine activities. For example, the Norse people believed lightning was caused by the god Thor throwing his hammer through the heavens. Some societies believed that weather was either a blessing or a punishment from the gods, depending on whether or not it helped the crops that had been planted. Likewise, some peoples believed that sickness and disease were caused by evil spirits or were inflicted upon humans by the gods as punishment for sin or disobedience. On the other hand, in the 2003 book *Modern Physics and Ancient Faith*, author Stephen Barr notes that the Judeo-Christian Bible did not attempt to explain natural occurrences in terms of spiritual entities but rather focused on the relationship between God and humans.

A cursory review of trends in several areas of human development over the last few centuries indicates an ever-increasing belief that more and more of the universe and human existence can be explained as natural phenomena, while there is less and less need for spiritual explanations. In the 1500s and 1600s, scientists began to discover in earnest that natural phenomena could be explained in terms of physical forces that are described by scientific laws. They were able to formulate scientific laws that applied to material bodies ranging in size from large planets to small objects in laboratories. In the 1600s, the laws of motion and gravitation were discovered by scientists such as Isaac Newton and Johannes Kepler. In the next several hundred years, the laws governing thermodynamics, electricity, magnetism, and light (electromagnetic waves) were likewise formulated. In the field of biology, scientists such as Louis Pasteur discovered in the 1860s that bacteria and viruses cause disease.

In the early 1900s, important discoveries were made about the nature of the atom and subatomic particles, and the theory of quantum mechanics was developed. During the same period, Albert Einstein developed the theory of special relativity to explain the nature of constant motion, especially at velocities near the speed of light. He hypothesized the inherent equivalence of matter and energy as described by his famous equation $(E=MC^2)$, which is explained in the footnote below.[1] He developed the theory of general relativity which revolutionized the understanding of the nature of acceleration and gravity.

Scientists discovered evidence that all of the physical forces of nature, some of which were once believed to be spiritual phenomena, could now be explained by scientific laws. Based on scientific evidence, lightning is no longer believed to be caused by the hammer of Thor but is believed to be an electrical discharge.

---

1. The equation states the basic equivalence of Energy (E) and Mass (M), $E = M$. Energy and matter are different forms of the same thing. The "C" in the equation has a constant value (the speed of light) which is used as a factor to convert units of mass into equivalent units of energy.

Based on scientific evidence, disease is no longer believed to be caused by evil spirits but is believed to be caused by bacteria and viruses. Based on scientific evidence, the sun is no longer believed to be a fire pulled across the sky by a chariot of the gods but is believed to be a thermonuclear reaction of gases. Around it, the Earth and other planets revolve according to the laws of gravity with the apparent motion of the sun caused by the Earth spinning on its axis. Based on scientific evidence, weather is no longer believed to be a blessing or a punishment from the gods. Rather, weather is believed to be primarily driven by energy from the sun in a complex interaction of uneven heating of the Earth's surface, water vapor, other gases in the atmosphere, cloud formation, ocean currents, the spinning of the Earth on its axis, the tilting of the Earth's axis, and the revolution of the Earth around the sun.

Many scientists and nonscientists alike no longer feel a need to believe in a spiritual world, or they reserve judgment until scientific evidence for such a world can be supplied. They want to see evidence that will stand up to the scientific method and be tested under the strict rules of scientific experimentation, which they believe to be the best method for determining "truth." Since all attempts to detect the spiritual world and spiritual beings using scientific instruments and techniques have failed, some scientists have discarded the spiritual world as being "unscientific" or even "antiscientific" and a carryover from the Dark Ages when myth and superstition ruled people's minds. Religion has been viewed by some as—at best—a needless waste of time and effort and—at worst—as an antagonist to scientific progress. Many scientists are aware of the persecution of fellow scientists as heretics several centuries ago, resulting in imprisonment and death for some scientists who supported theories that were claimed to contradict the official teachings of the church.

After the scientific awakening of the 1500s and 1600s, the need for a supernatural understanding of the world was discarded in a number of areas of human learning and discovery. New theories in economics, biology, philosophy, and psychiatry provided explanations of the world in purely natural terms.

In the area of economics, Karl Marx had no need of God and declared religion to be the opiate of the people. In 1848, he wrote *Manifesto of the Communist Party* which described all historical activity in terms of a theory of two opposing material forces (known as the doctrine of *dialectical materialism*) with the winning force developing the seeds of its own destruction. Finally, the proletariat (the working class) would rise and seize control of the means of production, after which the state would fade away and all competing forces would vanish.

In 1859, Charles Darwin, a naturalist, published *On the Origin of Species by Means of Natural Selection,* which provided a theory that all life on earth evolved from early life forms over billions of years in a process that relied upon nature forces. In 1953, Francis Crick and James Watson discovered the chemical structure of deoxyribonucleic acid ("DNA") and ribonucleic acid ("RNA") which provided the "blueprint for life" and explained the mechanism by which genetic information is passed on to offspring. It likewise provided the mechanism for explaining how organisms could mutate and evolve.

Friedrich Nietzsche, an existentialist philosopher, declared in the 1880s that "God is dead," meaning that people no longer believed in or had a need for God. Nearly a century later, the April 8, 1966, cover of *Time* magazine echoed Mr. Nietzsche with the question "Is God Dead?". Jean Paul Sartre, a philosopher, espoused in the mid-1900s his atheistic version of existentialism, the philosophy which describes aspects of "being human."

In the field of psychiatry, in the early part of the 1900s, Sigmund Freud disavowed any belief in a God or supernatural souls and laid the groundwork for modern psychiatry with a "scientific" theory of human development. He rejected the idea that mental disorders are caused by demons and evil spirits. His theory relied on understanding the human psyche in terms of unconscious feelings, governed by an id, an ego, and a superego.

As we proceed in this new millennium, what then is the current relationship between scientific discovery and the belief in human supernatural souls? Can the scientific and the spiritual world views be reconciled with each other? Or are they moving further apart as scientists are able to understand and explain more and more of the world in natural terms, with less and less need for any supernatural explanations? Will research ultimately enable scientists to explain how the matter that makes up the human brain can cause consciousness and a human mind?

Ironically, as described below, in spite of the historical inclination of many scientists to divorce science[2] from religion and the supernatural world, science actually provides solid, overwhelming evidence that leads to the logical conclusion that human beings have supernatural souls.

The Nature of Scientific Evidence

The scientific logic that leads to the conclusion that human beings have supernatural souls, different and distinct from their material bodies, is found in an understanding of the nature of science itself. Science is the investigation of the natural world and natural phenomena in order to explain observations and use a testable system to discover the laws that govern nature. Used in this sense, the natural "world" includes the entire universe: all matter, energy, and forces (including gravity, electromagnetism, and the strong and weak nuclear forces)[3] discovered by scientists over the last several hundred years.

As described in the chapter "Quantum Mechanics," the overwhelming scientific evidence is that all natural phenomena, including the movement and interaction of all matter and energy,

---

2. Throughout this book, the term "science" refers to the natural sciences. It does not include the "social sciences" which seek to discover the laws affecting the interactions of humans. Although there are certain common human characteristics and interactions that can be identified by the social sciences, humans have free will, which allows them to think and do things that are "free" and not subject to any laws.

3. These are the traditional four "fundamental" forces most of us learned in physics class. As described in the chapter "Quantum Mechanics," advances in scientific theory during the 1900s have combined the electromagnetic and weak nuclear forces.

can be understood in terms of scientific laws. Scientific laws allow scientists to predict the interactions of matter, energy, movement, and forces. Identical experiments produce identical results. This is the essence of the scientific method.

Scientific laws represent correlations between separate events. These correlations are also often referred to as cause-effect relationships, although some scientists would say that describing one event as a "cause" and another event as an "effect" requires a "belief" which is not strictly a part of science. Also, there are phenomena for which a "cause" has not been identified. These include such phenomena as the radioactive decay of atomic nuclei. Although no cause has yet been identified for radioactive decay, there is a well-defined decay rate depending on the type of matter present. One can therefore say that a correlation exists between the type of matter present and the rate of the radioactive decay of its atomic nuclei.

If one uses the "cause-effect" terminology, the big bang[4] beginning of the universe is the initial "cause" of all subsequent activity in the universe. All interactions of matter and energy following the big bang are due to prior causes or, as in the case of radioactivity, are due to the type of matter present. These cause-effect relationships and/or correlations have been discovered for natural phenomena even if due to the Heisenberg uncertainty principle[5] we are not able to define a cause-effect relationship that allows us to predict and measure the movements of individual atomic particles. The incredible technological developments of our modern society are based on this understanding.

Scientists and engineers study scientific laws to apply them in practical applications and get predictable results. A certain telephone is expected to ring 1,000 miles away every time a cer-

---

4. The "big bang" is the cosmic explosion theorized by scientists to have occurred at the beginning of the universe. Scientific evidence indicates that the big bang occurred about 15 billion years ago.

5. The Heisenberg uncertainty principle is described in the chapter "Quantum Mechanics."

tain number is dialed. A beam of a certain type of steel and of a certain size and thickness is expected to support weight of a certain amount. A radio tuned to a certain frequency is expected to pick up the same radio station day after day. A mixture of certain chemicals in specified amounts and conditions is expected to always produce the same reaction and products.

Science Is Not a Belief System

It is important to realize that science is not a belief system. Rather, as described above, science uses evidence to test hypotheses.[6] Thus, strictly speaking, it is scientifically incorrect to express a "belief" in any scientific hypothesis, theory, or law. For example, within the framework of science, a scientist should not say, "I believe there is gravity" or "I believe that evolution by natural selection has occurred." Rather, a scientist should more correctly say, "The evidence to date supports the hypothesis that gravity affects matter and energy" or "The evidence to date supports the hypothesis that evolution by natural selection has occurred."

This approach is based on the understanding that evidence might be discovered in the future that does not support the original hypothesis. If such evidence is found, then the original hypothesis must be modified to fit the evidence. For example, as described in the chapter "Quantum Mechanics," certain hypotheses of classical physics (which were once considered to be "laws" of physics)[7] had to be modified when new evidence concerning the characteristics of atomic particles was discovered at the beginning of the 1900s. Another famous modification to an established scientific "law" concerns the theory of general relativity developed by Albert Einstein in the early 1900s. Sir Isaac Newton had originally explained gravity as a force that causes two masses to attract each other. However,

---

6. An "hypothesis" is another name for a "theory."

7. After a significant period of time, an hypothesis or theory is considered to be a law if it is supported by a substantial amount of evidence and if there is no evidence that is inconsistent with the hypothesis.

according to the theory of general relativity, as confirmed by subsequent experimental results, it is more correct to explain gravity as something that warps time and space. Objects travel in straight lines through space warped by gravity such that they come together.[8] Thus, science does not "prove" any hypothesis or theory. It compares the available evidence with various hypotheses to identify the hypothesis that best fits the evidence.

I have used this scientific approach to identify the evidence which supports the hypothesis that humans have supernatural souls. There are two main bodies of evidence upon which I have relied:

- the evidence from decades of scientific experiments that indicate all natural phenomena (atoms, molecules, energy, and forces) interact according to laws; and
- the evidence that humans have free will.

As I explain in more detail below, the evidence that all natural phenomena interact according to laws implies that the source of free will cannot be explained as a natural phenomenon. This is because in order for free will to be truly free it must not be subject to laws. Thus, free will cannot be explained as a natural phenomenon that is subject to laws. By definition, a phenomenon that is not subject to the laws of nature is referred to as a supernatural phenomenon. This supernatural phenomenon which is the source of free will is generally known as a "soul."

Free will, if it exists, gives humans the ability to have belief systems and to choose to believe whatever they want. With free will people, including scientists, can choose, based on the quality and quantity of available evidence, whether or not to believe that evolution by natural selection has occurred, whether or not to believe there is gravity, and whether or not to believe humans have free will[9] and supernatural souls. However, these beliefs and

---

8. For an explanation of the theory of general relativity, see *The Elegant Universe* by Brian Greene (1999).

9. Of course, if humans do not have free will, they will not be able to freely choose whether or not to believe they have free will.

the expression of these beliefs are outside of science. Scientists and engineers rely on such beliefs to devise practical applications resulting in technological advancements. Likewise, everyone must decide whether or not the evidence for free will and supernatural souls can be relied upon as a basis for making life decisions.

Free will has important implications concerning science and belief systems in at least the following ways:

- Free will is used to judge the validity and quality of evidence.
- Free will is used to judge when the evidence is strong enough to warrant a belief that the hypothesis is true such that a person should rely upon it.
- Free will is used to determine which areas of scientific inquiry should be pursued and what types of experiments should be performed.

The above-described limitations of science and the scientific method vis-à-vis belief systems are not always fully appreciated, even by scientists. This is due in part to the cumbersome writing and speaking method that would be required to strictly reflect the ongoing nature of testing hypotheses and discovering new evidence. For example, it is much easier to say: "Force and acceleration are related by the following equation" than to say, "For all experiments conducted to date, the evidence supports the hypothesis that force and acceleration are related by the following equation." The use of the shorter declarative sentences such as those used in scientific text books is typically not a concern for scientific laws that have been extensively tested and found to be valid in all cases. Ideally, I think all text books should have a general disclaimer alerting the student or reader that all scientific "laws" are subject to further refinement and modification based on new discoveries and evidence.

Occasionally, in certain controversial settings, scientists may lapse into making statements such as: "Science shows that A causes B" rather than more correctly saying: "The evidence indicates that A and B are related in the following way." For example, there are scientific fields in which controversies exist due to the nature of available evidence and the implications that such evidence might have on human belief systems. An example of such controversy can be found surrounding the theory of evolution as discussed in more detail in the chapter "Biology." Another controversial issue is the hypothesis that human activity causes global warming. Controversies are heightened in many cases due to the implications that scientific discoveries and evidence have on public policy.

Any statements in this book that seem to claim that free will and supernatural souls exist reflect my belief system. I recognize that "scientifically" no hypothesis can be "proved." However, my beliefs are based on the evidence described herein.

At a very fundamental level, scientists have discovered the laws that govern how atoms and molecules interact.[10] In 1900, at the dawn of the twentieth century, the theory of quantum mechanics was first proposed by Max Planck to explain known atomic phenomena. In the 1920s, quantum mechanics evolved further by the development of the wave equation of Erwin Schrödinger. Quantum mechanics is now universally accepted as being a reliable predictor of the interaction of atomic particles. A more detailed description of the nature and theory of quantum mechanics is provided in the chapter "Quantum Mechanics."

Unfortunately, a number of scientists have used science's great success in discovering natural laws to reach an incorrect conclusion. Some scientists have reached the conclusion that anything that cannot be explained as a natural phenomenon does not exist. Some scientists claim that science teaches that there is no supernatural realm and that the supernatural realm is only superstition, myth, or fairy tale. However, science does not teach this. Science seeks to discover the laws that govern the natural world and has not directly detected the existence of the supernatural world. Lack of detection, however, does not imply lack of existence. For

---

10. As described in the chapter "Quantum Mechanics," there are theories concerning even smaller subatomic particles and waves.

example, electromagnetic phenomena were not detected by scientific experiments until relatively recently.

Some scientists have questioned how supernatural entities can interact with the natural world. This important question and concern is addressed in the chapter "The Soul-Brain Interface." One possible mechanism for this interaction is for human souls to have the capacity to create energy that can affect the human brain.

The notion that *anything that cannot be explained as a natural phenomenon does not exist* is only an *assumption*. As discussed in the chapter "Free Will or Not," some scientists have used this *assumption* to deny the existence of something that most people take for granted: human free will. Science is the search for the laws that govern the natural world. Suppose phenomenon "A" can be shown to exist but cannot be explained as an entity that is subject to natural laws. Based on all scientific evidence to date, which indicates that all natural phenomena are subject to natural laws, it would be logical to conclude that phenomenon "A" cannot be explained as a natural phenomenon and is best explained as something that is outside of or above nature. By definition, the realm that is outside of or above nature is the supernatural realm.

Since human free will cannot be explained as a phenomenon subject to natural laws, some scientists have reached the conclusion that free will does not exist. However, evidence that human free will *does* exist, leads to the conclusion that free will is best explained as a supernatural phenomenon.

Science has helped to clarify the distinction between the natural world and the supernatural world:

- **The natural world** includes all natural phenomena that are subject to the laws of physics and chemistry, including entities such as atoms, molecules, subatomic particles, and light.
- **The supernatural world**, on the other hand, includes entities that have free will and that are not governed by the laws of physics and chemistry.

Summarized below is a discussion as to why a natural world governed by scientific laws leads to the logical conclusion that human beings have supernatural souls.

THE EVIDENCE SUPPORTS THE HYPOTHESIS

The conclusion that human beings have immaterial, supernatural souls that are not subject to the laws of nature is summarized in the following logical explanation. If humans, including their brains, are only made out of matter (and possibly other natural phenomena) and the interaction of all matter and other natural phenomena are subject to scientific laws, then human free will is not possible. If, on the other hand, humans do have free will,[11] it follows that humans must be made of something more than just matter and other natural phenomena that are subject to scientific laws. This "something more" must be entities that are not subject to scientific laws and are what are normally referred to as supernatural souls. These supernatural souls are the sources of human free will. A more formal explanation is provided as follows.

To provide a logical basis for concluding that human beings have supernatural souls, I will start with the premise that if humans have free will, one of the two following hypotheses is correct but not both.

- **Hypothesis 1:** Humans have free will, which cannot be explained as a natural phenomenon. The source of human free will is a supernatural soul.
- **Hypothesis 2:** Humans have free will, which can be explained as a natural phenomenon. The source of human free will is not a supernatural soul.

If humans have free will, either **Hypothesis 1** or **Hypothesis 2** is correct, but **Hypothesis 1** and **Hypothesis 2** cannot both be true.

---

11. Evidence for human free will is provided in the chapter "Free Will Test."

Let us examine the evidence to see if it supports **Hypothesis 1** or **Hypothesis 2**. The overwhelming evidence is that all natural phenomena follow scientific laws. As described below, this evidence leads to the conclusion that free will cannot be explained as a natural phenomenon. Thus, the evidence does not support **Hypothesis 2** and we are left with **Hypothesis 1** as being the only logical explanation for free will.

Let us examine why the evidence does not support **Hypothesis 2**. We will assume that human beings do not have supernatural souls. If human beings do not have supernatural souls, then it follows that each entire human being is made up of the natural matter we call the human body. It is well accepted by the scientific community that all of the functions of the body are controlled by the central nervous system, which is made up of the brain, spinal cord, and nerves. Some of the bodily functions are controlled automatically by the autonomic nervous system. These include such functions as blinking, coughing to prevent choking, peristalsis (the movement of food through the digestive track), breathing, and blood circulation (by the beating of the heart), as well as the functions of a number of other bodily organs.

Some body functions are controlled directly by the brain on a conscious level. These include functions such as the movement of the muscles which control the arms, legs, neck, face, and torso. Some of the functions controlled by the autonomic nervous system can also be consciously controlled, to some extent, such as breathing, blinking, and coughing. There have even been some successful biofeedback experiments in recent years which indicate some people have the ability to consciously control their heart rate.[12] A more detailed description of the current understanding of how the brain works is provided in the chapter "The Soul-Brain Interface."

---

12. Chopra, Deepak (1993), pgs. 13, 33, and 84.

The brain and spinal cord are made up of nerves which transmit electrochemical signals to control all of the automatic and conscious functions of the body. One of the functions of the nervous system is to send electrochemical signals to the various muscles in the body. Many bodily functions are performed by muscles which move the parts of the body required to achieve the desired result. For example, breathing is caused by the contraction of the muscles in the diaphragm and chest wall. This contraction of muscles lifts the rib cage and expands the lungs causing them to draw in air. The heart is a muscle which contracts in a rhythmic fashion to pump blood through the body. The muscles are triggered to contract and then to release by electrochemical impulses that travel from the brain or the autonomic nervous system to the muscle.

CONSCIOUS MOVEMENTS

Conscious movements of the body are those that humans can decide whether or not to perform. For example, humans can decide whether or not to lift an arm. If a human decides to raise an arm, the brain sends out the appropriate signals over the nerves to the muscles in the shoulder and the arm. Then the muscles contract and the arm is raised. The nerves and the brain themselves are made up of molecules and atoms which move and function according to the laws of physics. Thus, the laws of physics determine how and when the electrochemical signals are transmitted.

What then is the source or starting point of any action of a human being? If there is no supernatural soul, then any bodily action, including an action resulting from a "conscious" decision, is the result of two conditions:

- the position and state of all of the atoms inside the body (including the atoms which make up the brain and nervous system); and
- any outside stimuli which are perceived by the body such as sounds, images, touching, temperature, etc.

Under this understanding, the brain is nothing more than a biological computer that processes incoming information (outside stimuli) and reacts based on the way the neural networks of the brain are connected (hardware) and the way the mental processes of the brain operate (software). At the most basic level, the atoms and molecules (and possibly other currently unknown natural phenomena) that make up the brain react according to the laws of physics and direct all of the brain's activity and the resulting actions of the body. Just as a computer cannot "decide" to act contrary to the way it has been programmed either initially or through "learning" while it has been subjected to external stimuli, so a brain that consists of atoms and molecules must act in the way it has been programmed through genetic codes and by its responses to external stimuli during the person's life.

Free Will

The one aspect normally associated with human beings that is missing from the above description is the concept of free will. If all human actions are the result of the interaction of the atoms and molecules that make up the brain and their reaction to outside stimuli, there is no room for the action of free will. Based on overwhelming scientific evidence, the interaction of atoms and molecules are governed by scientific laws. Free will, on the other hand, requires there to be something, such as a supernatural soul, that is not affected or controlled by physical forces and scientific laws. In order for free will to operate, there needs to be something that can freely choose to follow alternative decisions regardless of the outside stimuli. With free will, a person can choose one course of action one time and the opposite course of action the following time even if the circumstances are otherwise identical.

Without a supernatural soul, the atoms and molecules of the brain can only act in the ways they are forced to act by the laws of physics. Whatever the outside stimuli, the brain will react according to those laws. The reaction might be very complex and be based on a consideration of numerous factors identified over a lifetime and recorded in the brain. There is no scientific evidence, however, that the brain on its own can make a decision that is independent of the laws of physics.

FREE WILL VS. CONSCIOUSNESS

There have been a number of books written about consciousness over the years. In this book, however, I have decided to focus on free will rather than consciousness because consciousness is a much more difficult concept to analyze. Consciousness includes free will as well as other aspects such as awareness and feelings. While scientists and philosophers have debated for years whether or not consciousness can be fully explained by matter, I think it is clear that free will cannot be explained by matter that reacts according to the laws of physics.

NATURE AND NURTURE

Some people might not be bothered by the assertion that humans do not have free will. People that believe we are all shaped by the environment that surrounds us from the time we are born might be comfortable with a human being who is defined as the sum total of those outside influences imposed on a physical body. One school of thought has claimed that the genetic background of a person (nature) is the most important predictor of how someone will act. A separate school of thought holds that the environment (nurture) has the greatest influence. For the nurture theory, "environment" is defined as all outside influences (including parents, other family members, friends, and all people with whom the person comes in contact) from the beginning of life to the present. However, human actions cannot be totally explained by only nature and nurture. Some human actions are due to free will rejecting or overriding the influence of nature and nurture. Nature and nurture only influence how we feel in a certain circumstance. They do not force us to perform any specific action or to think in any certain manner.

A good example of how free will can override the effects of nature and nurture is that of a person who is genetically predisposed to alcoholism (nature) and is raised in an alcoholic family (nurture). In spite of the seemingly overwhelming influence of the forces of nature and nurture on this person, he or she can still use free will to choose whether or not to drink alcohol.[13] Although alcoholics might not be able to control their feelings, drinking alcohol requires many actions controlled by free will rather than by nature or nurture. These include such actions as taking money, going to a store, buying the alcohol, bringing it home, opening the bottle, pouring it into a glass, raising the glass, and drinking. The experience of millions of alcoholics who have stopped drinking alcohol indicates that in spite of the influences of nature and nurture, alcoholics can refuse to perform at least one of the above steps. Similarly, the actions of millions of other people addicted to drugs and chemicals indicate that people can use free will and choose to overcome their addiction.

The above discussion does not negate but rather confirms the enormous effort of the will that is often needed to overcome severe physical and psychological cravings to give in to the addiction. Neither does the above discussion disparage the efforts of those who have not yet been able to overcome their addiction.

No Free Will

What would be the consequences if human beings do not have free will? While philosophers have pondered this question for years, the practical consequences, if taken to their logical conclusion, would not be acceptable to most people.[14] The most far-reaching effect of not having free will would be that we would not have control over any of our actions and would not even have control over our thoughts. Consider the practical ramifications if humans do not have free will:

- All thoughts and the resulting actions are determined by the status of the atoms and molecules in the brain at any point in time and the outside stimuli acting on the body through the senses.

---

13. Many alcoholics say that they rely on spiritual help to make their free will choice to reject alcohol.

14. Of course, if humans do not in fact have free will, then none of us can choose what he or she will or will not accept.

- The concept of "choice" would be meaningless since choice implies the freedom to follow one course of action or another. If the atoms and molecules in the brain are simply following the laws of physics, there is no free choice.

- Without any free will or the ability to freely choose, there is no innocence or guilt and no right or wrong since no one has the ability to change what the atoms and molecules in their brains are forcing them to do.

- If someone harms another human being, it just happens, no one is free to choose. If a person affected by violence feels angry or sad, it just happens. Whether or not the victim of violence forgives the attacker is not important because no one can really choose one way or the other. All human actions are determined by brains which are just a bunch of atoms and molecules which have to act according to the laws of physics.

- No one can freely decide what laws to pass or what penalties to impose for crimes. All of the thoughts and actions of the people that enact laws or make rules are determined by the atoms and molecules in their brains.

- Striving hard to get an education, to do a good job, or to accomplish something are meaningless concepts. Either our brains are programmed to do these things or they are not. If someone has accomplished something, recognizing the person's efforts is meaningless because no one can make free decisions to exert such efforts.

- Humans cannot freely choose whether or not to encourage children to study and work hard. What adults do and say has already been determined by how they were raised.

- The stories of individual courage to overcome hardships and difficulties are not useful since no one can freely choose to do these things. Whether or not they happen is based on genetic coding and the stimuli people experienced throughout their lives. Part of their "stimuli" may include interactions with parents, friends, teachers, and other supporters, but those people also do not choose how to act or what to say; they are only reacting to their own coding and programming.

- If there is no actual free will, it would be truly ironic, considering the current public controversies over many areas involving choice such as funding options to enable children to attend nonpublic schools (school choice); the freedom to kill unborn human life (reproductive choice); the freedom to kill oneself or assist in voluntary suicide (the death with dignity choice); freedom to engage in prostitution, conceive children outside of marriage, or use drugs (lifestyle choices); the freedom to choose the supplier of telephone, cable television, electricity, or natural gas service (consumer choice); the freedom to choose the supplier of health care (health-care choice); the freedom to choose leaders and law-makers (voter choice); the freedom to choose to eat or to avoid eating foods high in fats and cholesterol (dietary choice); and the freedom to choose how and whether to invest money for retirement (investor choice).

- No one can truly take credit for making a scientific discovery (or for doing anything) because no one can make the required decisions or "choose" to think or study or act.

- We cannot freely choose to change what we think or believe because our brains are just atoms and molecules that follow the laws of physics. Any thought that we are thinking at any point in time is due to the status of the atoms and molecules in our brains, which is due to our genetic coding and all the stimuli we have encountered over a lifetime.

- We do not have the free will to choose whether or not to lift a finger or an arm. Such things are determined by our brains (which are composed of atoms and molecules that have to follow the laws of physics).

- The concepts of "truth," "wisdom," and "democracy" are meaningless since they imply that humans have the ability to freely choose.
- Life has no more meaning than one event following another based on the interaction of atoms and molecules (including the atoms and molecules that make up human bodies and brains) following the laws of physics. As the physicist Steven Weinberg says: "The more the universe seems comprehensible, the more it also seems pointless."[15]

Limits of Science

Most people stand in awe at the power of science and what can be created following scientific principles:

- instant communication using radios, television, and telecommunication;
- nuclear bombs that can destroy millions of people with a single explosion;
- nuclear power plants that can change matter into energy and provide electricity for millions of people;
- huge buildings that can rise a thousand feet above the ground and provide space for thousands of people to work and live;
- machines that can fly in the air and spacecraft that can take humans and equipment to the moon, other planets, and the far reaches of the solar system;
- computers that can perform complex tasks or repetitive and mundane tasks and multiply the creativity of human minds;
- medicines and vaccines that can wipe out deadly diseases; and
- Artificial organs and other body parts that can extend life or improve the physical condition of people.

---

15. Weinberg (1977), pg. 154.

In spite of all these wonderful and truly amazing accomplishments, there is one thing science cannot do and will never be able to do: science cannot provide an explanation of how a completely material human being, whose atoms and molecules must follow the laws of chemistry and physics, can have a free will and do simple things such as make free choices about everyday life or even choices about what he or she thinks. In spite of the absurdities that result if humans do not have free will, some scientists, in maintaining their intellectual honesty,[16] have gone so far as to question the existence of free will. They do so because the existence of free will would require that humans be made up of more than matter and other natural phenomena that follow the laws of physics. These scientists think that the existence of a spiritual world contradicts the basic premises of science.[17] However, as described previously, this is a misunderstanding of the nature of science.

Likewise, it is not valid to say that the inability to explain free will in terms of matter and forces is only due to the current state of brain science and that science may someday be able to explain how the brain allows for free will without a supernatural force. If the human brain is made up of matter and other natural phenomena that are subject only to the laws of science (regardless of what those laws may be or even whether or not we fully understand what those laws are) and is not acted upon by a supernatural force, science will never be able to explain free will.

QUANTUM RESPONSE

A number of scientists have recognized the inconsistency between free will and brains that are made up of atoms and molecules that have to follow scientific laws. However, they do not appear to

---

16. Of course, if they are correct and humans do not have free will, then honesty is a meaningless concept and they are only thinking and saying that there is no free will because they are forced to do so by the atoms and molecules that make up their brains.

17. For examples of statements by scientists who question the existence of free will, see the chapter "Free Will or Not" and the chapter "Evolution."

want to follow such a premise to its logical absurd conclusions, such as those listed above. One of the most creative approaches in trying to overcome this dilemma is to look to the role of consciousness in quantum mechanical experiments and the Heisenberg uncertainty principle which is a part of the quantum theory of physics.[18]

The free will aspect of human consciousness has a role in quantum mechanical experiments because the human experimenter can choose which quantum mechanical aspect of nature to investigate. However, human free will must be outside of the natural realm in order for the human brain of the experimenter to be able to freely choose which aspect of nature to investigate. If human free will were not outside of the natural realm, then the brain of the experimenter would be governed by the quantum mechanical state of the atoms and molecules that make up the brain. For such a case, the experimenter would not be truly free to choose which aspect of nature to investigate.

According to the Heisenberg uncertainty principle, there is a defined limit as to how accurately we can simultaneously measure the position and velocity of atomic particles. The Heisenberg uncertainty principle is theorized to provide the flexibility in the movement of atomic particles to account for human free will. However, the Heisenberg uncertainty principle operates on the basis of the laws of probabilities. Thus, the Heisenberg uncertainty principle cannot explain human free will because human free will must not be subject to any scientific laws, even probabilistic laws.

Free Will and Game Rules

In his 1949 book *The Concept of Mind*, Professor Gilbert Ryle claimed that free will can be a natural phenomenon that follows rules because rules do not necessarily determine what happens. His gives as an example the rules of games, such as the game of chess. He noted that even though chess players follow the rules

---

18. The quantum theory of physics and how it is used to try to account for free will is explained in more detail in the chapter "Quantum Mechanics."

of chess, the specific moves made by the players are "free" and the outcome of the game is not determined by the rules. Professor Ryle's reasoning, however, is not valid. The rules or laws of nature are not the same kind of rules as the rules of a game developed by humans. The rules of a game are better described as limits within which actions can be performed. It is correct that the actual moves in a chess game are not determined by the rules. It is human free will that determines the specific square to which a chess object is moved. Thus, although free will (in a chess game) can function within the rules of chess, the free will moves are not determined by the rules. The chess player can also make a free will decision to not play according to the rules. It is not possible to predict beforehand any specific move by a player or any combination of moves. The scientific method, on the other hand, is used to develop rules which are used to explain evidence and predict future interactions. Moves made by players in games developed by humans are "free" because they do reflect free will decisions by humans. The fact that people with free will can choose to limit their actions within the confines of the rules of a game does not imply that the rules which govern the interaction of natural phenomena are similar to the rules of human games.

The laws or "rules" of nature are very different from the rules of games made by humans. Overwhelming scientific evidence indicates that atoms, molecules, and other natural phenomena do not choose how they will move or interact. Using the laws of nature, it is possible to predict very accurately how matter and energy will interact. This is described in more detail in the chapter "Quantum Mechanics."

EMERGING PROPERTIES

Another approach used to try to explain the ability of matter to produce human minds is known as the "emerging properties" approach. This is part of the theory of evolution. Under this theory, consciousness and the human mind are properties that emerged or evolved from matter just like other biological properties, such as breathing, digestion, metabolism, etc. The 1987 book *Darwin and the Emergence of Evolutionary Theories of Mind and Behavior* by Robert J. Richards describes how the theory of mind as an evolutionary property developed in the years after Darwin's 1859 book *On the Origin of Species*. Subsequently, the concept appeared in publications such as the 1926 book *Holism and Evolution* by Jan Christian Smuts, the 1950 book *Cell and Psyche: The Biology of Purpose* by Edmund Sinnott, and the 1955 book *The Phenomenon of Man* by Pierre Teilhard de Chardin. In more recent times, the concept that the human mind is an emerging biological property has been hypothesized by Professor John Searle, Professor Daniel Dennett,[19] and many other philosophers and scientists.[20]

There may be some basis for accepting the premise that natural selection would favor a characteristic such as "consciousness" that makes animals more aware of their surroundings and thus better able to survive. The theory of emerging properties, however, is not valid when applied to free will. Biological properties can emerge from matter because biological properties do follow the laws of physics. Conversely, free will cannot emerge from matter because matter follows the laws of physics and free will must be outside the laws of physics in order for humans to be able to make free choices. As described in the chapter "Biology," Charles Darwin himself and Professor William Provine of Cornell University have recognized that there is no rational basis for free will being explained as a natural biological mechanism.

Logical Conclusion: Hypothesis 2 Is Not Correct

The above evidence indicates that free will cannot be explained as a natural phenomenon that follows laws and, thus, if humans have free will, **Hypothesis 2** is not correct. This leads to the conclusion that **Hypothesis 1** is correct and supernatural souls are

---

19. John Searle is professor of philosophy at the University of California and Daniel Dennett is professor of philosophy at Tufts University. See Searle (1992), Searle (1997), Dennett (1984) and Dennett (1991).

20. This concept is discussed in more detail in the chapter "Biology."

needed to explain human free will. As described in the appendix to this chapter, a simple logical syllogism can also be used to indicate why it is logical to believe in the existence of supernatural souls.

It is logical that a scientific explanation of human existence that recognizes human free will must also accept the existence of a supernatural soul. The only other alternative is to theorize that the atoms and molecules in human brains do not follow the laws of physics. This alternative envisions a "natural" dualism of mind and body under which the human mind emerges from the material world but is not subject to the laws of physics and is the basis for human free will. This theory is known as "nonreductive physicalism" and is described in more detail in the chapter "Quantum Mechanics." However, such an alternative is problematic:

- Nonreductive physicalism is completely nonscientific because there is not one shred of experimental evidence that would support the theory that some atoms and molecules do not follow the laws of physics. As described in numerous quotes by physicists listed in the chapter "Quantum Mechanics," the experimental results have been in 100% agreement with the hypothesis that all natural phenomena interact according to laws of science.

- The theory of nonreductive physicalism would have to explain how the atoms and molecules in the brain somehow "free" themselves from the laws of physics when other atoms and molecules continue to follow the laws of physics.

- The theory would have to explain how the atoms and molecules in the brain interact with each other and how they are able to make independent decisions. Most notably the theory would have to answer the questions: "who is in charge?" and "where does any decision start?". Do individual molecules have freedom to act, or are they somehow conscripted to operate together?

We thus come to a very interesting conclusion: belief in a supernatural soul is a logical scientific belief because it allows the matter in the brain to continue to act like matter and be subject to the laws of physics while the supernatural soul provides the source of free decisions and choices. Belief that supernatural souls do not exist is an illogical, nonscientific belief because it requires that the matter in the brain must have a "mind of its own" and must not follow the laws of physics so that free will can be explained.

Some scientists respond to the suggestion that there is a supernatural world with a "wait and see" attitude. They want scientific evidence before they accept something they cannot see. Normally, scientific evidence means performing experiments under controlled conditions and achieving the same result each time. This is known as the scientific method. If humans were only made out of matter, according to the scientific method, they would have to respond the same way each time they are faced with identical conditions. The evidence for the existence of the supernatural soul uses the scientific method by recognizing that human free will means that there is some nonmaterial free will agent that allows a human to choose to respond differently even if the conditions are the same. The chapters "Free Will or Not" and "The Soul-Brain Interface" also describe the evidence that humans with obsessive-compulsive disorder (OCD) can use their wills to produce a mental force that modifies the material brain and stops or reduces OCD behavior.

The scientific method does not require that something be physically detected before its existence is accepted by scientists. No one has ever seen or measured the electron wave around an atom defined by the Schrödinger wave equation.[21] Yet the description of an electron as a wave around an atom is firmly accepted by physicists because it completely explains the characteristics that can be measured. As described in the chapter "Quantum Mechanics," no one has seen virtual particles, but their existence

---

21. This concept is described in more detail in the chapter "Quantum Mechanics."

is accepted because scientists can detect the effect that the virtual particles have. Likewise, there is evidence for supernatural souls because we can see the effect that they have on human bodies. The existence of supernatural souls explains how humans can have free will.

How does free will work?

If atoms and molecules in the brain must follow the laws of physics, how do humans exercise free will? A logical answer is that:

- the soul is a "free agent" that is not subject to the laws of chemistry and physics, and is the source of willful decisions and free will;
- the soul and the brain interact with each other;
- the soul has the ability to produce a mental force that can affect the atoms and molecules in the brain such that they transmit the decisions made by the soul to the appropriate part of the body; and
- the soul has the ability to detect signals given off by the brain, thereby receiving information gathered by the body's senses.

A discussion of the possible mechanisms for the mental force produced by the soul and the operation of free will and brain-soul couplings based on current knowledge about how the brain functions is contained in the chapter "The Soul-Brain Interface." Although we might never be able to determine exactly how the soul interacts with the matter of the brain, as described in that chapter, there are at least plausible explanations based on the laws of physics. On the other hand, it is impossible to develop an explanation of how matter that follows the laws of science can on its own produce free will without a supernatural source.

The Supernatural Soul

The above discussion indicates that the soul is a supernatural entity. Before the reader thinks that I am going to start talking about ghosts and goblins, I must again explain that the prefix "super" simply means "over" or "above" or "outside of." Thus the term "supernatural" means "over or above or outside of nature." The supernatural soul is over or above or outside of the natural body. The soul controls or directs the body similar to a supervisor or overseer[22] that controls or directs an employee. As discussed above, the soul must be outside of or above the laws of nature so that it can be a free agent that allows humans to have free will.

The term "*super*natural" is in a very real sense similar to the term "*super*ego" which is a scientific term used in Sigmund Freud's theory of human psychology to describe that part of the human psyche that is on a higher moral plane and is "over" or "above" the ego. Dr. Freud may have been closer to describing a supernatural soul than he would have cared to admit.

Need for a Free Will Test

The above discussion is based on the assumption that humans have free will. Some people will take the position that not having free will is an acceptable conclusion.

My response to that is this:

- if humans do not have free will, then no one can choose whether or not to believe that humans have free will and it is meaningless to talk about it (other than the fact that if we do talk about it, it is because we have somehow been programmed to talk about it and then it is meaningful only in the sense that all movement is meaningful because it follows the laws of physics); and
- if humans do not have free will, then we must be willing[23] (an inherent contradiction) to accept the consequences described above (that is, we cannot control

---

22. Note that "super" means "over" and "visor" means "seer." The word "visor" comes from the same root as "vision" which means "sight."

23. Note how often during the average day we make comments which imply we as humans have the power to choose something or another. Is this freedom to choose just an illusion?

what we think and that good and evil, laws and crimes, love and hate, and choice are meaningless concepts).

Nevertheless, some scientists and philosophers have come to the conclusion that humans do not have free wills (although I do not think they fully appreciate or admit the implications of that belief). With this in mind, it would be beneficial to devise a scientific test that provides evidence that humans do in fact have free will and can make willful choices. Such a test is described below.

THE FREE WILL TEST

As described in more detail in the chapter "Free Will Test," it is possible to devise a test that provides evidence that humans do have the ability to make free, "willful" choices. The basis of the test is that there are certain occurrences that have a predetermined probability of happening, based on the laws of statistics. A human being can interact with the test to force an outcome that would not occur randomly and the results for which there is no reasonable explanation, other than free will. Based on the evidence from the free will test, it is logical to conclude that humans have free will.

The ability of OCD patients to decide to exert effort to change the OCD behavior that otherwise would have continued to occur absent such a decision is additional evidence that humans have free will.

THE WORTH OF HUMANS

Free will and spiritual souls give us worth as human beings. Free will and spiritual souls also make the inalienable rights bestowed on us by our Creator become meaningful, such as the rights listed in the United States Declaration of Independence, in the United Nations Declaration of Human Rights, and in the constitutions of many countries. Those human rights include life, liberty, the pursuit of happiness, and the security of person. Without supernatural souls, humans cannot make free choices, and these stated rights would become meaningless because we would not be able to freely choose whether or not to pursue or enjoy them. With free will, humans can choose to help people and create value that will bring happiness to others. It therefore follows that each human life has value and that people must have the rights of life, liberty, the pursuit of happiness, and the security of person so that they can be free to choose good, to help others, and to create value for themselves and for others.

Note that all human people have free will and can make free choices regardless of gender, culture, religion, or national origin. Thus, democratic institutions that protect the rights of each person represent a form of social contract that is in keeping with the characteristics of all people. Likewise, free market economies provide people with the maximum amount of choice which is in keeping with the nature of humans who can make free choices.

Morality and Meaning

The ability of humans to make free choices provides solid evidence that good and evil exist and that humans were created with free will so that they could choose good or evil. As described above, if free will did not exist, people would not be able to choose their actions, and the concepts of good and evil would be meaningless. Learning what is good and what is evil is only the first step. Each human must then decide whether to choose to accept God and pursue good or reject God and pursue evil.

Likewise, as described by Viktor Frankl in his 1959 book *Man's Search for Meaning*, life is meaningful because humans have free will, which allows them to make meaningful choices and to make decisions that have real consequences. Without free will, people would just go through the motions of life and not be able to decide whether or not they will choose to accept God, pursue good, help others, and create value.

School Curriculum

The above discussion that free will cannot be explained as a natural phenomenon is based on scientific evidence and logic. It is not based on religious faith or religious dogma. Thus it should be possible (and I think desirable) to discuss in any classroom and school in the world the fact that science cannot explain human free will as a natural phenomenon that must follow the laws of science.

MIRACLES HAPPEN

By definition, a miracle is something that cannot be explained by natural occurrences that follow the laws of physics. I believe it is logical to conclude that the creation of each human being includes the joining of a supernatural soul with a material body, which cannot be explained as a completely natural occurrence. It involves a supernatural explanation and thus, in a very real sense, the creation of each human being is a miracle.

The ability of humans to make free will decisions is a profound human characteristic. It cannot be explained by a natural explanation that is based on a being that is only made of matter. It requires a supernatural explanation. In that sense, each free will decision made by a human being is a miracle. I believe that a realization and appreciation of that fact will produce in each of us a feeling of awe and wonder. We should always be aware that human beings are first and foremost spiritual beings. We should always be aware of how special and miraculous it is to be able to make free decisions and choices.

ETERNAL SOULS

We have no way of determining scientifically what happens in the spiritual world. However, in my opinion, it is logical to believe that the creation of a soul and the joining together of a soul and a body require an almighty, omnipotent supernatural being. It likewise seems logical to me that if an omnipotent supernatural being creates another supernatural being such as a human soul, it would be so that this new supernatural being would live forever. Science has evidence that the matter and energy of the universe cannot be destroyed by natural processes. This is known as the law of energy conservation. On this basis, I think it is logical to believe that our supernatural souls, which are more valuable than matter and energy, should likewise not be capable of being destroyed and will live forever. Knowing that an almighty God can create a supernatural soul and join it with a material body should make it easier to believe that the same God can resurrect bodies at the end of the world and reunite them with their souls.

Practical Results

Once we have used our free will to accept the evidence that leads to the conclusion that humans have supernatural souls, we must deal with it on a practical basis as follows:

- The conclusion that humans have supernatural souls should affect our basic principles and how we live our lives.
- It should affect how we treat other human beings, all of whom have supernatural souls and the inalienable rights of life, liberty, the pursuit of happiness, and the security of person.
- We should be filled with a feeling and sense of awe and wonder that humans have bodies that can operate both automatically and with an infinite gradation of purposeful control by the soul. Consider the "simple" act of walking, which most people do without thinking. There is an infinite gradation in the degree to which we become aware of how we are walking and the degree to which we make purposeful, willful choices about how we walk, dance, run, hop, skip, jump, play sports, etc. A similar comparison can be made for all other areas of human activity. Science can only explain willful choice and control by recognizing the existence of a supernatural soul.

CHAPTER CONCLUSIONS

1. Human beings have free will that allows them to make free choices.

2. Human brains that are only made up of atoms and molecules (and possibly other natural phenomena) that are subject to the laws of physics cannot on their own make free choices.

3. The conclusion that material brains cannot on their own make free choices is based on the scientific evidence that the interaction of all matter and energy follows the laws of physics and is not based on the laws of physics that happen to be known at any point in time. As long as the interaction of matter and energy follows the laws of physics, no refinement of the laws of physics due to a future scientific discovery will affect or change this conclusion.

4. Free will cannot emerge or evolve from matter because matter is subject to the laws of physics. Matter on its own is not free to make free choices because matter follows the laws of physics.

5. The inability of science to explain free will as a natural phenomenon does not mean that free will does not exist.

6. Science only seeks to discover the laws that govern the natural world. Science does not claim there is not a supernatural world.

7. The only explanation for free will that does not contradict the evidence currently available is that each human being has a supernatural soul that is closely integrated with, but also separate from, the material body. A supernatural soul provides the free agent necessary for free will.

8. There are plausible, scientific explanations as to how a supernatural soul could interact with a material brain.

There are no scientific explanations as to how a material brain made up of atoms and molecules that follow the laws of physics can be the source of free will.

9. Science is not a belief system. Science is a system of using evidence to test hypotheses or theories. There have been instances in the history of science when new evidence has required hypotheses to be revised. All science text books should include a disclaimer that all theories are subject to further modification based on the discovery of new evidence.

10. People use their free will to choose what they believe. Scientists, as well as all people, should base their beliefs on the quality and quantity of available evidence. There is overwhelming evidence that all natural phenomena are subject to the laws of physics. Thus, there is overwhelming evidence that the atoms and molecules that make up human brains (which are natural phenomena) are not the source of human free will.

11. Scientists believe in phenomena they cannot see as long as it explains the phenomena they can see and there is evidence to support that relationship. Supernatural souls explain how humans can have free wills. The existence of supernatural souls is supported by the overwhelming evidence that all natural phenomena are subject to the laws of physics.

12. If humans do not have free wills and cannot make free choices, the concepts of right and wrong, good and evil, love and hate are meaningless.

13. If humans do not have free will, humans do not even have the ability to make simple choices such as what to say or think, even though they think they do have the ability to make free will choices. If humans do not have free will, we are living an existence of total absurdity.

14. If humans do not have free will, then it is not possible to freely choose which scientific principles or laws we will believe and rely upon.

15. The concept that human beings have free will that cannot be explained as a natural phenomenon that is subject to scientific laws is based on scientific evidence and logic and should be taught and fully debated in all schools.

16. Having supernatural souls gives us worth as human beings and allows us to pursue and enjoy our God-given rights of life, liberty, the pursuit of happiness, and the security of person.

17. The concept that each human has a supernatural soul is solidly based on scientific principles and evidence because it allows matter in each human brain to continue to follow the laws of physics.

18. Believing that humans do not have supernatural souls is not based on scientific evidence. This is because such a belief requires that matter in each human brain must have a "mind of its own" and not follow the laws of physics so that it can explain human free will.

19. It is logical to believe that the creation of each human being is a miracle that includes the joining of a supernatural soul with a material body to enable each human being to have free will. It is logical to conclude that this requires the power of an almighty supernatural being.

20. It is logical to believe that a God that is powerful enough to create a supernatural soul and join it to a human body is also powerful enough to resurrect the body at the end of the world and rejoin it with its supernatural soul.

21. Believing in free will and supernatural souls should have a practical effect on our attitudes and how we live our lives.

Appendix To Introduction and Basic Premise Chapter

Logical Syllogism Supporting the Existence of Supernatural Souls

The following syllogism (logical argument)[24] also provides another way of logically concluding that supernatural souls exist.

It is first necessary to explain the relationship between a conditional statement (CS) and its contrapositive (CP). If a conditional statement (CS) is true, then the contrapositive (CP) is also true.

Conditional statement (CS):   If A then B.

Contrapositive (CP):   If not B, then not A.

A simple example will explain the above logical reasoning. Suppose that:

A =   Rain

B =   Wet grass.

Not A =   No rain

Not B =   No wet grass

Then we could say:

Conditional Statement (CS):   If it rains (A), then the grass will be wet (B).

If the conditional statement (CS) is true, we know the contrapositive statement (CP) is also true.

Contra-positive (CP):   If the grass is not wet (not B), then it did not rain (not A).

We know that contrapositive statement (CP) is true because if it did rain, then the grass would be wet. If the grass is not wet, we know it did not rain.

---

24. Students take note that this Appendix is based on logic I learned in high school.

A similar syllogism can be used to logically conclude that supernatural souls exist. For this syllogism, we will use the following parameters:

A = a human being is only made up of matter[25]

B = a human being does not have free will

Not A = a human being is made up of more than matter

Not B = a human being does have free will

The syllogism, then, would be:

Conditional Statement (CS): If a human being is only made up of matter (A), then a human being does not have free will (B).

Contra-positive (CP): If a human being does have free will (not B), then a human being is made up of more than matter (not A).

Based on the evidence that matter always follows the laws of physics, matter alone does not provide for an explanation of free will and we can logically conclude that the conditional statement (CS) is true. If the conditional statement (CS) is true, then the contrapositive statement (CP) is true. The free will test described in the chapter "Free Will Test" provides evidence that humans do have free will. Thus, based on the validity of the contrapositive statement, we can logically conclude that humans are made up of more than matter. This something "more" is something that is nonmaterial and is not a natural phenomenon. It is something supernatural: a supernatural soul.

---

25. For this example, matter refers to all natural phenomena that are subject to the laws of physics.

"The greatest gift which humanity has received is free choice. It is true that we are limited in our use of free choice. But the little free choice we have is such a great gift and is potentially worth so much, that for this itself, life is worthwhile living."

—*Isaac Bashevis Singer*

"We have to believe in free will. We have no choice."

—*Isaac Bashevis Singer*

Chapter Two
FREE WILL OR NOT

The logical conclusion (based on scientific evidence[1]) that humans have supernatural souls is dependent on the existence of free will. Based on their everyday experience, most people, I would venture, believe that humans have free will and are able to make free choices. But how has free will been treated historically by teachers, philosophers, and scientists?

In Greek mythology, the three goddesses that controlled the lives of humans were known as the Fates. There are many stories in Greek mythology in which people suffered the consequences of performing acts that they were "fated" to perform. In English, the word "fate" has come to mean a person's predetermined destiny which that person is unable to change.

In the Christian world, the Apostle Paul started the controversy by his references to what has been interpreted by some to be predestination.[2] In the 400s, Aurelius Augustine, Bishop of Hippo (a.k.a. St. Augustine), had debates with Pelagius over the nature of human free will, faith, human interaction with the grace of God, and the salvation of the soul. Note that the Apostle Paul and Augustine both believed that humans have spiritual souls and

---

1. The scientific evidence that all natural phenomena interact according to the laws of physics.

2. See, for example, chapters 8 and 9 of the Apostle Paul's letter to the Romans and chapter 1 of his letter to the Ephesians. See also the Gospel of Matthew 24:22, Acts 13:48, 1 Peter 1:2 and Revelations 17:8.

that there is a spiritual God. To the extent Paul and Augustine believed that human free will is limited, they believed it is limited by the power of a spiritual God. Their belief in a limitation on human free will was not based on a claim that humans do not have supernatural souls as some philosophers and scientists claim. The writings of Paul and Augustine influenced Martin Luther in the 1500s, when he challenged the teachings of some Church authorities, leading to the Reformation. The concepts of free will and predestination have been discussed from religious and spiritual perspectives for centuries, and the controversies will not be resolved here.[3]

In the 1600s, philosopher René Descartes reasoned that the very act of doubting his own existence indicated that he did exist because there must be something that is doing the doubting. This is encapsulated in his famous conclusion:

"I think, therefore I am."[4]

Mr. Descartes believed that humans have free will and that the act of doubting was evidence of that free will. He viewed this ability of doubting to be evidence that humans are essentially "thinking beings." He concluded that humans are made up of two different types of things: a material body that is extended over space (*res extensa* in Latin) and a supernatural soul that is the "thinking thing" (*res cogitans* in Latin). This is known as Cartesian dualism. Philosophers and scientists struggled with the concept of how a supernatural soul could interact with a material body. This is why many scientists and philosophers reject the concept of a supernatural soul. Asking how a supernatural soul can interact with a material body is a valid question. This question is addressed in the chapter "The Soul-Brain Interface."

---

3. The details of this and subsequent debates throughout Christian history are described in the 1997 book *Willing to Believe: the Controversy over Free Will* by R. C. Sproul. See also *On the Freedom of a Christian* (1520) and *The Bondage of the Will* (1525) by Martin Luther.

4. In Latin, this is "*Cogito, ergo sum.*"

The preceding saying by Descartes can be paraphrased for the purposes of this book as follows:

I can make free choices, therefore I am a supernatural soul.

Although this reformulation is not as succinct as the original statement, it expresses the essence of Descartes' saying by helping to identify the critical aspects of "thinking" and "being." The important aspect of "thinking" relative to Descartes' logic is the ability to doubt, which is a free choice. The important aspect of "being" is existence as a supernatural "thinking thing." This understanding of "being" is a better explanation of what it means to be human than existence as a collection of atoms and molecules. Viewing a human as merely a collection of atoms and molecules does not capture the essence of what it means to be human. This is because atoms and molecules interact according to the laws of chemistry and physics, leaving no room for free choices.

In the early 1800s, scientist and mathematician Pierre Laplace recognized that all matter in the universe moves according to the laws of physics. He proclaimed that all of history is determined by the laws of physics and that all of history could thus be predicted if we could only know the position and momentum of each bit of matter in the universe (including the matter that makes up all human brains) at one point in time.[5] He was not aware, however, of the Heisenberg uncertainty principle discovered in the early 1900s that says the exact position and momentum of elementary particles cannot be known simultaneously. Even though the movement and interaction of all matter follows the laws of physics, the Heisenberg uncertainty principle would prevent us, even in principle, from being able to determine simultaneously the position and momentum of all matter and thus also prevent us from being able to predict history.

In recent times, people that believe all human actions are determined by the laws of science have become known as "deter-

---

5. Thornton (1989), pg. 70, as excerpted from *Analytical Theory of Probability* (1820).

minists," even if the laws of science are the probabilistic laws of quantum mechanics.[6] The concepts of fate and determinism bring into question the existence of human free will. If everything, including the atoms and molecules in human brains, moves and interacts according to the laws of science, how can there be free will?

FREE WILL DEFINES WHAT A HUMAN IS

Free will is the most basic characteristic defining what a human being is. It is the very essence of being human. Without free will, a human being would not be what we normally think of as a human being. It is impossible to overstate the importance of free will to humans. Determining whether or not free will exists and, if so, identifying the source of free will are the most important questions addressed by philosophy and science. In fact, most of philosophical thought throughout history has revolved in one way or another around these questions.

Without free will, life becomes absurd (as described in the chapter "Introduction and Basic Premise"). Without free will, all human actions and even human thoughts result from matter following the laws of physics, and life has no meaning. Without free will, all human actions and thoughts are reactions to external stimuli based on genetic coding (nature) and upbringing (nurture). Without free will, we cannot freely decide whether or not we will be morally good or bad, whether or not we will work hard, which person, if any, we will marry, what career we will pursue, or what religion we will believe. Without free will, we cannot even freely decide what we will eat for breakfast, what clothes we will wear, or what we think. Without free will, we cannot freely decide whether or not we believe in free will.

During the Age of Reason in the 1600s and 1700s (also known as the Enlightenment), some scientists and philosophers

---

6. For a discussion of the laws of quantum mechanics, see the chapter "Quantum Mechanics."

identified the ability to reason as the characteristic that separates humans from other animals. Free will is an inherent and important component of what is meant by the "ability to reason." The ability to reason is more than just being able to execute a logical operation like a computer. The ability to reason includes the ability to determine what is and what is not logical. Free will is even necessary to decide when the available evidence supports a scientific theory or to determine whether or not a logical argument is valid. Free will allows humans to decide whether or not to choose how to act based on what is logical and "reasonable." Ironically, although the Age of Reason championed human reason and the scientific method (which needs human free will to function), some scientists doubt the existence of human free will based on what they think is a scientific perspective.

Of course, choices often do not represent the need to employ reason and make a logical decision. Rather, some choices are just preferences with no right or wrong answer. Some people choose apples and some choose bananas. Some people choose dates and some choose figs.

Free will brings meaning to life because it allows our decisions to become real decisions that have consequences, and it allows us to choose how we will live our lives. With free will, each human being becomes a unique creature that can decide what he or she will and will not do, regardless of genetic background, upbringing, the surrounding circumstances, and the external stimuli. That is the essence of being able to make free will choices.

Or so we all think. How can we be sure what we do is not just a complex response to our surroundings determined by our genetic coding? Maybe we are just another group of animals that happen to be very intelligent. While you are sitting there trying to decide if you have enough free will to be able to decide whether or not you have a free will,[7] consider the following:

---

7. I take comfort from the knowledge that if humans do in fact have free will, then my basic premise is correct. If humans do not have free will, then I had no real choice as to whether or not to write this book.

- In 1957 the New York University Institute of Philosophy hosted an entire conference on the question of whether or not humans have free will.[8]
- In 1958 and 1961, after eight years of work, the Institute for Philosophical Research published two volumes of *The Idea of Freedom*. In the second volume, over three hundred pages were devoted to delineating and clarifying the controversy between the determinists and those who believe in free will and freedom of choice.[9]
- Several prominent scientists, philosophers, and psychiatrists have made statements either claiming or implying that free will is an illusion.
- Since the 1600s, scientists and philosophers have debated, based on scientific principles, whether or not humans have free will. For an excellent discussion of the debate, see the 1989 book *Do We Have Free Will?* by Mark Thornton.
- There are numerous books on free will listed at the end of this chapter and in the bibliography of this book.
- In the 1999 book, *The Volitional Brain,* edited by Benjamin Libet, a number of scientists and philosophers addressed the question of whether or not science can explain how a brain made out of matter can be volitional and can make free will choices.

As described by physicist Henry Stapp in *The Volitional Brain:*

A controversy is raging today about the power of our minds. Intuitively, we know that our conscious thoughts can guide our actions. Yet the chief philosophies of our time proclaim, in the name of science, that we are mechanical systems

---

8. See Hook (1958).

9. Adler (1985), pg. 151.

governed, fundamentally, entirely by impersonal laws that operate at the level of our microscopic constituents.

The question of the nature of the relationship between conscious thoughts and physical actions is called the mind-body problem.[10]

As discussed in the chapter "Introduction and Basic Premise," those who say that science denies the existence of free will (as suggested in the above excerpt) because free will cannot be explained as a natural phenomenon are misinterpreting the nature of science. Also, it would be my guess that most people intuitively believe that humans have the ability to make free choices and would be surprised that "a controversy is raging" as to whether or not humans have free will.

Why is there so much controversy over the existence of human free will that the above-described activities have occurred? Does it not seem absurd that if a creature does not have free will, it would (or even could) think and write so much about free will? How is it possible that some very educated philosophers, scientists, and psychiatrists could come to the conclusion that humans do not have free will in spite of overwhelming, everyday, common sense evidence to the contrary? These philosophers, scientists, and psychiatrists have come to this conclusion because they have first believed that humans are only made up of matter and other natural phenomena. From such premise it follows logically that all human thoughts and actions are controlled and are not free because all evidence to date indicates that matter and other natural phenomena move and react according to scientific laws.

They have declined, however, to seriously consider the alternative hypothesis: that humans are made up of more than just natural phenomena. They have misinterpreted the nature of science. Science does not require or teach that there is not a supernatural world. Science does not seek to investigate the supernatural

10. Libet (1999), pg. 143.

world. Science only seeks to determine the laws that govern the natural world.

The philosophers at the 1957 conference who believed in free will were not able to explain the existence of free will based on a "body only" human. Nevertheless, they could not gather up enough courage to utter the word "soul" and then try to explain why they thought it is not a valid explanation of free will. The proceedings for the 1957 conference do not include any reference to the word "soul" or any discussion of Socrates'[11] concept of a supernatural psyche (Greek word for "soul") that controls the actions of human bodies. In his 1971 book, *Socrates*, Professor W.K.C. Guthrie[12] described how Socrates considered the human body to be a tool that is controlled by the psyche. He described how in the dialogue *First Alcibiades* (written by Plato) Socrates expresses his belief that:

> . . . *in speaking of a man we mean something different from his body—that, in fact, which makes use of the body as its instrument. There is nothing that this can be except the psyche, which uses and controls . . . the body.*[13]

With the long tradition of Western philosophers who believed in supernatural souls, it is amazing to me that a group of philosophers gathering to discuss determinism and free will would not at least consider the effects of a psyche or soul, even if the concept of something supernatural would have ultimately been discounted. It is also amazing that some philosophers and scientists could believe that all human actions and thoughts are controlled by the laws of chemistry and physics, recognizing that it makes everything absurd. I think that the philosophers did not consider the possibility of a supernatural soul or psyche because they were

---

11. Socrates was a Greek philosopher who lived in the 400s B.C.

12. Professor W.K.C. Guthrie, F.B.A. was master of Downing College and Lawrence professor of ancient philosophy at the University of Cambridge.

13. Guthrie (1971), page 152.

afraid to appear "nonscientific" or silly, as if they were to propose that they believed in myths and fairy tales.

The scientists who contributed to the 1999 book *The Volitional Brain* were likewise unable to explain how a material brain could be the source of human free will. Henry Stapp[14] and David Hodgson[15] proposed in *The Volitional Brain* that free will can be explained by the interaction of "consciousness" with quantum mechanical measurement. As explained by Dr. Stapp:

> *The solution [to the mind-brain problem] hinges not on quantum randomness, but rather on the dynamical effects within quantum theory of the intention and attention of the observer.*[16]

I do not agree with the above explanation of Dr. Stapp. As described in the chapter "Quantum Mechanics," the results of quantum mechanical experiments are very predictable, leaving no room for the source of free will to be a quantum mechanical phenomenon. The ability of humans to choose which aspect of reality to investigate with a given quantum mechanical experiment indicates that there is something outside of quantum mechanics that is the source of the free will choice.

The fact that the free will of the experimenter or observer is involved in making quantum mechanical measurements does not imply that the source of free will is a natural phenomenon. On the contrary, if the source of free will were a natural phenomenon it would be subject to the laws of chemistry and physics and the scientist would not be able to "freely" choose which aspect of nature to investigate.

---

14. Henry Stapp is a physicist at Lawrence Berkeley National Laboratory, University of California.

15. David Hodgson is author of the 1991 book *The Mind Matters: Consciousness and Choice in a Quantum World*. He is a justice on the Supreme Court of New South Wales, Australia.

16. Libet (1999), pg. 146 in Dr. Henry Stapp's article "Attention, Intention, and Will in Quantum Physics."

Likewise, the photons or electrons in a quantum mechanical experiment do not "decide" whether they will demonstrate their wave nature or their particle nature. That is chosen by the experimenter. For the experimenter to have the ability to freely choose which aspect of nature to investigate, the experimenter must ultimately be controlled by something that is not subject to the laws of chemistry and physics.

As described by Dr. Stapp in a paper entitled "A Bell-type Theorem without Hidden Variables," experimenters must be able to make free choices:[17]

> *For the purposes of understanding and applying quantum theory, the choice of which experiment is to be performed in a certain space-time region can be treated as an independent free variable localized in that region. [Niels] Bohr[18] repeatedly stressed the freedom of the experimenter to choose between alternative possible options. . . . This 'free choice' assumption is important because it allows the causal part of cause-and-effect relationships to be identified: it allows the choices made by experimenters to be considered to be causes. [Emphasis in the original.]*

Note that Dr. Stapp identifies free choice as being an independent cause. As described in the chapter "Introduction and Basic Premise," one of most basic concepts of science is that, in nature, the interaction of all matter and energy is subject to cause-effect relationships (or, as some scientists would say, certain interactions are correlated). By identifying free choice as an independent cause, Dr. Stapp is identifying free choice as being outside of nature and therefore a supernatural phenomenon. For free choice to be an independent cause it must not be caused by

---

17. The full text of this paper can be found on Dr. Stapp's website at *http://www-physics.lbl.gov/~stapp/stappfiles.html*

18. Niels Bohr was one of the original developers of some of the principles of quantum mechanics and of an interpretation that became known as the Copenhagen interpretation of quantum mechanics.

a natural phenomenon that is subject to prior causes. Although God is often described as the ultimate uncaused Cause, human free will allows humans to exert Godlike power as they are the agent of each independent, uncaused free will choice that they make.

There is nothing in the theory of quantum mechanics whereby atoms and molecules (such as those that make up human brains) can decide the probabilities of interaction they will exhibit as they interact with other atoms and molecules. There is nothing in the theory of quantum mechanics whereby atoms and molecules (such as those that make up human brains) can decide to go against the probabilities of interaction with other atoms and molecules as determined by the principles of quantum mechanics. An outside force, however, might be able to affect the probabilities of interaction. Thus, quantum mechanics might be the mechanism for implementing a free will decision made by an entity outside of the natural world. However, the quantum mechanical characteristics of the natural world are not sufficient in themselves to be the source of free will.

In an April 19, 2001, email to University of California Philosophy Professor John Searle, Dr. Stapp provides further clarification as to the nature of "freedom" and "choice" in quantum mechanics. The first type of choice is one that nature makes and one that is subject to statistical rules. An example of this type of "choice" made by nature is the choice as to where an electron is found around an atom. The location of an electron, when it is measured, is based on statistical or probabilistic rules. Since the exact location cannot be determined before the measurement, there is "freedom" as to where the electron might be. This is the same understanding of "freedom" used by Professor Danah Zohar as described in the chapter "Quantum Mechanics." This understanding of freedom is associated with the collapse of the wave function. This understanding of freedom is not a description of free will.

The second type of "choice" is one made by the experimenter regarding which aspect of nature to view. An example of this type of choice is the choice to view either the particle nature or wave nature of an electron. Dr. Stapp explains his position to Professor Searle:

> *The other point of discussion between us concerned "freedom": I identified the freedom of choice that I was talking about as a choice that is not determined by the laws of quantum mechanics, as they are NOW KNOWN: i.e. the laws formulated by [John von Neumann].*
>
> *Quantum theory has TWO choices of this kind:*
>
> 1) *Nature's choice of which of the two "outcomes" will appear. (This choice is subject to statistical rules, and in that sense is NOT "completely free.")*
>
> 2) *The choice of which question to put to Nature.*
>
> *It is the second choice that is connected to "Free Will."*
>
> *If NOTHING AT ALL fixes such a free choice, I would call this choice a "whimsical" choice. But that is not the only possibility. For the known laws of [Quantum Theory] may be only part of the story: there may [be] other laws that fill the dynamical gap.*
>
> *The nature of these further possible laws is still unknown. They might tie the "free will" choice to the brain in some nonlocal way. Even the possibility of some "spiritual" influence cannot be logically ruled out.*
>
> *I believe that a rational moral philosophy cannot allow one's "free will" to be controlled by any of the following three possibilities:*
>
> 1) *local mechanical micro-process,*
>
> 2) *random statistical choices,*
>
> 3) *pure whimsy.*

The importance of [John von Neumann's Quantum Theory] is that this framework encompasses what science knows about Nature, but leaves open, as a rational possibility, that our moral free choices are determined by none of the three processes listed above, but are the outcomes of, for example, a nonlocal process of self-examination and evaluation.[19] *[Capitalized words in the original.]*

Note that Dr. Stapp realizes that a process that is subject to rules, even statistical rules, is not "completely free" and is not what we mean when we talk about free will. Rather, free will must be completely free and not subject to any rules. He identifies the choice that the experimenter makes regarding which question to put to nature as a free will choice. He also admits that the source of this choice could be a spiritual source. Dr. Stapp correctly concludes that anything controlled by rules or laws is not "completely free." Based on this logic, free will must not be controlled by anything in the natural realm because overwhelming scientific evidence indicates that everything in the natural realm is subject to rules and laws. For free will to be "completely free" it must be caused by a supernatural phenomenon that is not subject to any rules or laws. As he suggests, the process for implementing a free choice might be instantaneous across time and space (which is referred to as "nonlocal"), but the source cannot be from the natural realm.

Free Will and Mental Force

In their fascinating 2002 book *The Mind and The Brain: Neuroplasticity and the Power of Mental Force*, Dr. Jeffrey M. Schwartz, M.D. and Sharon Begley describe the scientific evidence for free will and its ability to affect a material brain based on the treat-

---

19. April 19, 2001, email from Henry Stapp to John Searle. The complete text can be found at the website of Dr. Henry Stapp: *http://www-physics.lbl.gov/~stapp/stappfiles.html* under the heading "Emergence and Free-Will (Exchange with Searle, April, 19, 2001)."

ment of people who have suffered strokes, obsessive-compulsive disorder (OCD), Tourette's syndrome, and depression. The results of the treatments indicate that people are able to make willful choices to use directed mental force to change their brain patterns. This is the "smoking gun" evidence that a supernatural soul that is the source of free choices can produce the mental force needed to affect a material brain. Or as Dr. Henry Stapp says, it is "prima facie evidence" that the human will can affect the material brain.[20] This is described in more detail in the chapter "The Soul-Brain Interface."

In *The Mind and The Brain*, the authors appear to try to explain this "mental force" as a natural phenomenon in terms of the interaction of "consciousness" with quantum mechanical measurements. The authors explain that they have relied on Dr. Henry Stapp for their understanding of quantum mechanics and consider him to be a "virtual third coauthor."[21] However, as explained above in this chapter, in the chapter "Quantum Mechanics," and in the chapter "The Soul-Brain Interface," the source of free will cannot be explained as a natural quantum mechanical phenomenon. Although quantum mechanics might describe the *mechanism* by which a free will decision is implemented, the *source* of free will must be something that is not subject to the laws of physics.

At a November 15, 2003, symposium,[22] Dr. Schwartz explained that OCD patients must use effort to overcome the quantum mechanical probabilities inherent in the brain patterns that would otherwise result in the patients continuing with their OCD behavior. The important role of "mental effort" is also described in *The Mind and The Brain*:

---

20. Schwartz (2002), pg. 297.

21. Schwartz (2002), Acknowledgements, pg. xi.

22. The symposium Darwin, Design, and Democracy IV was presented at the University of Minnesota by Intelligent Design Network, Inc. and The MacLaurin Institute.

> *. . . I argued that the undeniable role of effort and the pos-*
> *sibility of an associated mental force to explain the observed*
> *changes in the OCD circuit suggest a mechanism by which the*
> *mind might affect—indeed, in a very real sense, reclaim—the*
> *brain.*[23]

The effort required to change brain patterns cannot be explained as a natural phenomenon of the brain because there is overwhelming evidence that atoms and molecules must follow the laws of quantum mechanics and cannot "decide" on their own to change the quantum mechanical probabilities of interaction.

In *The Mind and The Brain*, Dr. Schwartz reflected Dr. Stapp's understanding that quantum mechanical experiments and inter-actions require that a question be posed by the experimenter as to which aspect of nature is to be investigated:

> *Without some way of specifying what the question is, the*
> *quantum process seizes up like a stuck gear and grinds to a*
> *halt.*[24]

It is important to realize that the choice of which aspect of nature to investigate comes from outside of nature. As Dr. Schwartz explains, the choice is *not* prescribed by the quantum mechanical laws of physics:

> *Formulating that question requires a choice about which*
> *aspect of nature is to be probed, about what sort of information*
> *one wishes to know. Critically, in quantum physics, this choice*
> *is free: in other words, no physical law prescribes which facet of*
> *nature is to be observed.*[25]

Dr. Schwartz explains that it is the freedom to choose, that comes from outside the quantum mechanical states of the brain,

---

23. Schwartz (2002), pg. 295.

24. Schwartz (2002), pg. 282. This is nearly a verbatim quote from Henry Stapp's article "Attention, Intention, and Will in Quantum Physics" in Libet (1999), pg. 154.

25. Schwartz (2002), pg. 294.

that provides the ability to affect the brain. Without that outside force, the quantum mechanical brain states just continue to "evolve deterministically:"

> *. . . the entire brain of an observer can be described by a quantum state that represents all of the various possibilities of all of its material constituents. That brain state evolves deterministically until a conscious observation occurs. . . . Because the observer's only freedom is the choice of which question to pose . . . it is here that the mind of the observer has a chance to affect the dynamics of the brain.*[26]

The above excerpts imply that there is something outside the atoms and molecules of the brain that is the source of the choice by OCD patients to make an effort to change their behavior. This change of behavior requires a choice to *go against the quantum mechanical probabilities* of interaction that would continue to occur in the atoms and molecules of the brain if the OCD patient did not exert such effort.[27]

The authors of *The Mind and The Brain* view the mind as a phenomenon that emerges naturally from the human brain. They believe the characteristics and behaviors of the mind cannot be explained in terms of the sum of its parts and thus cannot be wholly explained by the brain.[28] Note, however, that it is only an *assumption* that the human mind is an emergent phenomenon capable of making free decisions. This assumption is not supported by the evidence of decades of quantum mechanical experiments, which indicate that all natural phenomena interact according to the theory of quantum mechanics. There is no part of the theory of quantum mechanics whereby a property of free will (which allows choices that do not follow any laws) can emerge from natural phenomena that follow the laws of chemistry and physics.

---

26. Schwartz (2002), pg. 285.

27. This is discussed in more detail in the chapter "Quantum Mechanics."

28. Schwartz (2002), pg. 350.

The authors of *The Mind and The Brain* base their claim that the mind is an emergent phenomenon on experiments which purport to show that there are instantaneous interactions between separated particles of matter and energy in the universe. This instantaneous interaction is a phenomenon called "nonlocality" as explained in the 1999 book *The Nonlocal Universe: The New Physics and Matters of the Mind* by science historian Robert Nadeau and physicist Menas Kafatos.[29] The experiments that purport to provide evidence that indicates matter can interact faster than a signal traveling at the speed of light are controversial.[30] However, even assuming that all matter in the universe is closely tied together by these instantaneous interactions, this does not discredit Cartesian dualism and the need for an uncaused, independent, supernatural source for human free will. The authors of *The Nonlocal Universe* only *assume* that the human mind is a natural phenomenon. They do not explain how these instantaneous interactions of matter and energy that follow the laws of physics give rise to free choices by humans.

The authors of *The Mind and The Brain* claim "Cartesian dualism was a disaster for moral philosophy, setting in motion a process that ultimately reduced human beings to automatons."[31] I disagree that Cartesian dualism is to blame for this "moral disaster." To the extent humans are considered to be automatons, it is due not to *acceptance* of Cartesian dualism but is due to *rejection* by scientists of Cartesian dualism. Most scientists reject the supernatural aspects of Cartesian dualism and the ability of

---

29. Schwartz (2002), pg. 349.

30. See, for example, the article "Does Bell's Inequality Principle Rule Out Local Theories of Quantum Mechanics?" at *http://math.ucr.edu/home/baez/physics/Quantum/bellsinequality.html*. Note that Albert Einstein claimed that the speed of the transfer of information from one point to another cannot exceed the speed of light. Be aware also, however, that Albert Einstein was never completely comfortable with quantum mechanics and that his theory of general relativity is not completely consistent with quantum mechanics.

31. Schwartz (2002), pg. 373.

a supernatural entity to interact with a material entity. However, the logic of Cartesian dualism was valid in the 1600s and is valid yet today. Brain matter that functions according to the laws of physics (whether they are the laws of classical mechanics or quantum mechanics)[32] cannot on its own be the source of free will. René Descartes only lacked knowledge of the quantum mechanical mechanism that could explain the soul-brain interaction. Dr. Schwartz's research and treatment of patients with OCD using free will as a mental force supports rather than refutes Cartesian dualism. The "mental force" identified by Dr. Schwartz is perfectly consistent with and equivalent to the "*res cogitans*" or "thinking entity" of Cartesian dualism. Since the time of René Descartes, scientists have rejected Cartesian dualism because there was no explanation as to how a supernatural entity could interact with and change a material brain. Quantum mechanics provides the required mechanism, and the work of Dr. Schwartz provides the scientific evidence that a supernatural soul (which Dr. Schwartz calls the mind or human consciousness) can produce a mental force that modifies brain metabolism and circuits. Dr. Schwartz explains that brain matter on its own cannot produce this mental force. Human consciousness cannot come about simply as a byproduct or "epiphenomenon" of the physical activity of the brain. As explained by Dr. Schwartz, an outside mental force is needed to change brain activity and the remodeling of brain circuitry:

> As noted in Chapter 1, this philosophical position, known nowadays as epiphenomenalism, views conscious experience as nothing more special than the result of physical activity in the brain, as rain is the result of air pressure, wind, and cloud conditions in the atmosphere. Epiphenomenalism is a perfectly respectable, mainstream neurobiological stance. But it denies that the awareness of a conscious experience can alter the physical brain activity that gives rise to it. As a result, it seemed to

---

32. See the chapter "Quantum Mechanics" for a description of classical mechanics and quantum mechanics.

me, epiphenomenalism fails woefully to account for the results I was getting: namely, that a change in the valuation a person ascribes to a bunch of those electrochemical signals can not only alter them in the moment but lead to such enduring changes in cerebral metabolic activity that the brain's circuits are essentially remodeled. That, of course, is what PET [positron emission tomography] scans of OCD patients showed.[33]

For the "mental force" to be able to reflect free will choices, the mental force must originate in something that is not subject to the laws of physics. It is logical to conclude that this mental force originates in a supernatural soul that uses the mechanism of quantum mechanics via the human brain to make quantum choices and collapse wave functions. A recognition and acceptance of the validity of Cartesian dualism (which operates via these mental forces) restores human spirituality, free will, and morality.

The authors of *The Brain and The Mind* also note that Cartesian dualism served science by relegating the material world to science and the supernatural world to the Church.[34] I agree that this is an advantage of recognizing that humans have a supernatural soul. However, I reject any implication that this was a result contrived by René Descartes. As described in the following pages, the statements by modern scientists who deny the existence of human free will confirm the logic of Mr. Descartes.

Some modern scientists (including scientists that understand quantum mechanics) have recognized that the atoms and molecules of the human brain must follow the laws of chemistry and physics, thereby leaving no room for explaining free will as a natural, biological phenomenon. The statements listed below were made by several prominent scientists in some popular books written in the 1980s and 1990s. They imply or claim outright that free will is just an illusion. Rather than revise their belief that

---

33. Schwartz (2002), pg. 292.
34. Schwartz (2002), pg. 33.

humans are only made out of matter, they instead accepted the conclusion that humans do not have free will.

I am thankful that a number of brilliant scientists have written so honestly about their belief that humans do not have free will. Such writings help demonstrate that a belief that humans are only made up of matter (and possibly other natural phenomena) leads to the conclusion that humans do not have free will. Alternatively, if you freely choose to believe humans have free will (assuming of course you have a free will so that you can freely choose what you believe), then the logical conclusion is that humans are made up of more than matter and do have supernatural souls.

None of the books on free will I have reviewed have been able to explain how a being made up only of matter can have free will. Some authors will recognize that humans have a "mind" but only as a "natural" phenomenon. They do not recognize or are unwilling to admit that free will cannot be explained as a natural phenomenon that follows the laws of phyics.

PSYCHIATRISTS

The belief of many psychiatrists that free will is an illusion is described by psychiatrist Rollo May in his 1969 book *Love and Will*:

> *Is will an illusion? Many psychologists and psychothera-pists, from Freud down, have argued that it is. The terms "will power" and "free will," so necessary in the vocabulary of our fathers, have all but dropped completely out of any contemporary, sophisticated discussion; or the words are used in derision. People go to therapists to find substitutes for their lost will: to learn how to get the "unconscious"to direct their lives, or to learn the latest conditioning technique to enable them to behave, or to use new drugs to release some motive for living.*[35]

---

35. May (1969), pg. 15.

STEPHEN HAWKING

In his popular book *A Brief History of Time*,[36] the famous astrophysicist, Professor Stephen Hawking, describes free will as an illusion:

> *Of course, one could say that free will is an illusion anyway. If there really is a complete unified theory*[37] *that governs everything, it presumably also determines your actions. But it does so in a way that is impossible to calculate for an organism that is as complicated as a human being. The reason we say that humans have free will is because we can't predict what they will do.*[38]

Professor Hawking is saying that he does not believe humans have free will. Nor can they freely decide what they will do or think. He believes that all human thoughts and actions are determined by the laws of physics. All the atoms and molecules in the universe, including those that make up human brains, are governed by the laws of physics. In addition, physicists are attempting to consolidate all physical laws into a complete unified theory. Note that his belief that humans are made up only of matter leads directly to the conclusion, based on logical reasoning, that humans do not have free will.

He recognizes, however, that in the human body there are trillions of nuclear particles whose interactions are exceedingly complex. Due to the number of particles and the complexity of the interactions, we will never be able to develop an accurate mathematical model of the interactions. Thus, we will never have a mathematical model that would enable us to predict the future

---

36. The 1996 edition of *A Brief History of Time* indicates that 9 million copies have been sold. Professor Hawking holds the chair as Lucasian professor of mathematics at the University of Cambridge. This is the same position held by Sir Isaac Newton, the famous scientist that discovered the laws of motion in the 1600s.

37. The complete unified theory of physics is discussed in the chapter "Quantum Mechanics."

38. Hawking (1988) 1996 ed., pg. 167.

actions of those particles and thus the future actions of the entire human body. The problem of developing an accurate mathematical model becomes even more incredibly difficult when we realize that each human is affected by his surroundings and other human beings. We would have to include in the mathematical model not only the human being in question, but also all the atomic particles that make up his or her surroundings and every human with whom he or she comes in contact. Professor Hawking believes that all the atomic particles act in accordance with some complex mathematical model but the interactions of all the particles are so complex that it is not practical to develop the mathematical model. Thus, Professor Hawking says: "we can't predict what [humans] will do."

Professor Hawking is well aware of the laws of quantum physics. He understands that human experimenters can choose whether a quantum mechanical experiment will demonstrate the wave aspect or the particle aspect of the phenomenon being studied. He knows that, based on the Heisenberg uncertainty principle, the exact position and momentum of individual particles cannot be simultaneously determined. He realizes, however, that, contrary to what some people have proposed, neither the involvement of human consciousness in a quantum mechanical experiment nor the inability to determine the exact position and momentum of individual particles provides a mechanism for particles to have a mind of their own. Thus, he believes that quantum physics cannot explain free will as a natural phenomenon and considers free will to be an illusion.

There are other noted physicists that held the same view as Professor Hawking. As related in the 2002 book *The Mind and The Brain:*

> In 1931, Einstein had declared it "man's illusion that he [is] acting according to his own free will."[39]

---

39. Schwartz (2002), pg. 299.

The ability of human consciousness to choose which aspect of reality to investigate with a quantum mechanical experiment implies that the choice must come from outside the experimenter's brain. Otherwise, the choice as to which aspect of reality to investigate would not be free but would be determined by the quantum mechanical state and associated probabilities of the atoms and molecules making up the brain neurons of the experimenter.

The fact that the Heisenberg uncertainty principle results in a causal indeterminancy relative to the movement of atomic particles has no bearing on whether or not humans have free will. It is an entirely different question. As explained by Professor Mortimer Adler in his 1985 book *Ten Philosophical Mistakes*:

> *The causal indeterminancy involved in certain scientific formulations, especially quantum mechanics, simply bears no resemblance to the causal indeterminancy involved in freedom of choice.*
>
> *What the determinists who deny freedom of choice on the grounds stated above fail to understand is that the exponents of free choice place the action of the will outside the domain of the physical phenomena studied by science.*[40]

In other words, Professor Adler is saying that human free will cannot be explained as a natural phenomenon and is "outside the domain of the physical phenomena studied by science." Free will allows a person to make a definite choice. Free will is not the same thing as the causal indeterminancy due to the quantum mechanical characteristics of atoms and molecules.

It is important at this point to realize that Professor Hawking has confused his reason (Reason 1 below) as to why we are not able to predict human actions with the correct reason (Reason 2):

- Reason 1 (Professor Hawking): According to Professor Hawking, it is theoretically possible to develop a mathematical model that would predict the actions

---

40. Adler (1985), pg. 149.

of humans. Thus, humans *do not* have free will. However, the number of particles involved is so enormous and the interactions of those particles are so complex that it is not practical to describe them by a mathematical model.

- Reason 2 (correct reason): It is *not* theoretically possible to develop a mathematical model that would predict the actions of humans because humans *do* have free will and can change what they want to do "at will."

So, you might ask, how do we know that Reason 2 is the correct reason we cannot predict human actions? How do we provide evidence that humans do have free will and are not just a complex arrangement of atoms and molecules, with the "illusion" of having free will? There are several approaches. One approach is the free will test described in the chapter "Free Will Test." Another approach is to examine the evidence of free choices made by OCD patients to go against the quantum mechanical probabilities of their brain patterns that would have occurred absent their free choices.[41] A third approach is to examine the noncomputational aspects of human thought as described in the chapter "Math."

As Professor Hawking describes, it is not practical to use a mathematical model to predict the future action of a human being (this would be a prediction from outside the person). It is my hypothesis, however, that a supernatural soul enables a person to predict what he or she will do (a prediction from inside the person). Thus a person can declare what he or she will do and then do it. The free will test described in the next chapter provides an example of how a person can decide what the results of a test will be and then make it happen.

---

41. See the chapter "The Soul-Brain Interface."

Free Will and the Scientific Method

Professor Hawking also recognizes that a lack of free will would present an enormous problem for the scientific method and the study of science. If humans do not have free will, how can they be "free" to discover scientific truths, to decide when the evidence is valid, to decide when the evidence explains the hypothesis or theory, and to make logical deductions? As explained by Professor Hawking in *A Brief History of Time*:

> *Now, if you believe that the universe is not arbitrary, but is governed by definite laws, you ultimately have to combine the partial theories into a complete unified theory that will describe everything in the universe. But there is a fundamental paradox in the search for such a complete unified theory. The ideas about scientific theories outlined above assume we are rational beings who are free to observe the universe as we want and to draw logical deductions from what we see. In such a scheme it is reasonable to suppose that we might progress ever closer toward the laws that govern our universe. Yet if there really is a complete unified theory, it would also presumably determine our actions. And so the theory itself would determine the outcome of our search for it! And why should it determine that we come to the right conclusions from the evidence? Might it not equally well determine that we draw the wrong conclusion? Or no conclusion at all?*
>
> *The only answer that I can give to this problem is based on Darwin's principle of natural selection. The idea is that in any population of self-reproducing organisms, there will be variation in the genetic material and upbringing that different individuals have. These differences will mean that some individuals are better able than others to draw the right conclusions about the world around them and to act accordingly. These individuals will be more likely to survive and reproduce and so their pattern of behavior and thought will come to dominate. It has certainly been true in the past that what*

we call intelligence and scientific discovery has conveyed a survival advantage. It is not so clear that this is still the case: our scientific discoveries may well destroy us all, and even if they don't, a complete unified theory may not make much difference to our chances of survival. However, provided the universe has evolved in a regular way, we might expect that the reasoning abilities that natural selection has given us would be valid also in our search for a complete unified theory, and so would not lead us to the wrong conclusion.[42]

Professor Hawking's contention is that free will is not needed to explain our ability to discover scientific laws because natural selection (the explanatory mechanism of evolution as conceptualized by Charles Darwin) has resulted in the human race having reasoning abilities that enable it to recognize useful scientific laws. The principles of evolution help humans to develop ever more technologically advanced civilizations that help humans to survive.[43] Professor Hawking's reliance, however, on Darwin's theory of natural selection to explain how science can progress without humans having free will is flawed for several reasons:

- Most importantly, Professor Hawking does not provide evidence that humans do not have free will. He only *assumes* humans do not have free will and provides an apparent alternative explanation of how humans can learn scientific laws. Moreover, other statements by Professor Hawking indicate that he does in fact believe that humans can make truly free choices that are not just the illusion of free choices.

- The *assumption* that humans do not have free will is based on the *assumption* that humans are only made up of matter that follows the laws of physics. Scien-

---

42. Hawking, (1988) 1996 ed., pgs. 12–13.

43. Note that if Professor Hawking is correct that humans do not have free will, then the great discoveries he has made are not due to his decision to study and work hard. They only happened because of evolutionary forces.

tific evidence that humans do have free will requires scientists to discard the theory that humans are only made up of matter.

- There is no aspect of Professor Hawking's explanation that would invalidate the free will test described in the next chapter that provides evidence that humans do have free will.
- As Professor Hawking admits, according to the theory of natural selection, the scientific theories that are "correct" will not necessarily prevail, but only those scientific theories that enable humans to survive and procreate will ultimately prevail.
- Based on the theory of evolution, the people that have survived and have offspring have presumably done so because they have characteristics that give them a survival advantage. This, however, does not provide any basis for deciding which people are using "correct" and "valid" logic to make scientific discoveries or to decide which scientific discoveries are valid. Presumably the "scientific" statements made by all people who have survived and have offspring are equally valid.
- If humans do not have free will, then there is not even any ability to choose on a personal basis which scientific laws we believe are valid. Presumably, if a person (including any scientist) is predisposed to believe certain laws are valid, he or she will believe they are valid and if not so predisposed, he or she will not believe they are valid.
- The "reasoning abilities" needed for making scientific discoveries require free will decisions by the humans using them. A scientist has to be able to freely choose whether or not a certain scientific theory explains all of the scientific evidence. "Reason" alone without free will is not adequate to make scientific discoveries.
- In science, the questions that are asked and the areas of research that are pursued are probably at least as

important as being able to make logical deductions about the results of experiments. A scientist needs free will to be able to decide which questions to ask and which areas of research to pursue.

• As described in the chapter "Math," the "reasoning abilities" that Professor Hawking attributes to natural selection include noncomputational aspects of mathematical reasoning that cannot be explained as being produced by a biological computer, such as the brain, that must operate according to the laws of physics and chemistry.

The previous statements by Professor Hawking concerning free will and natural selection indicate the intellectual struggle that must go on if you believe that humans do not have free will, while at the same time you also witness the thousands of free decisions that humans make every day. There are occasional statements by Professor Hawking in *A Brief History of Time* that indicate he does believe in human free will. For example, he asks "What should you do when you find you have made a mistake...?"[44] If humans do not have free will, this is an absurd question. Without free will, there is no possibility of changing your actions to do what you "should" do rather than what you "should not" do. Without free will, you will just do what your nature and nurture direct you to do.

Professor Hawking also remarks in his book that he visited the Vatican for a conference on cosmology and he states that he "had no desire to share the fate of Galileo"[45] who was subject to house arrest by Church officials in the 1600s on account of his scientific beliefs. However, if no one has a free will, would he agree then that anything that happened to anyone in the past, including Galileo, was not the "fault" of anyone and that the Church officials did not have any real choice in deciding to

44. Hawking (1988), 1996 ed., pg. 155.
45. Hawking (1988) 1996 ed., pg. 120.

arrest Galileo? According to his logic, they only did what the laws of physics made their material bodies do. I am sure he does not really believe that, and neither do I. Each of us is responsible for our own actions and thoughts. Likewise, today's Church officials are not responsible for the actions of past officials.

I appreciate the fact that many scientists do not believe in supernatural souls because it is not possible to directly detect a soul using scientific apparatus. However, Professor Hawking has himself provided us with the observation that often things that exist can only be detected indirectly. As he explains:

> *The force-carrying particles exchanged between matter particles are said to be virtual particles because, unlike "real" particles, they cannot be directly detected by a particle detector. We know they exist, however, because they do have a measurable effect: they give rise to forces between matter particles.*[46]

Likewise, the existence of human free will can be explained by and is evidence for the existence of supernatural entities which can affect material particles and give humans free will.

ROGER PENROSE

On page 35 of *A Brief History of Time*, Professor Hawking indicates that he has collaborated with Professor Roger Penrose on certain scientific investigations. Professor Penrose addressed human consciousness in a popular 1994 scientific book entitled *Shadows of the Mind: A Search for the Missing Science of Consciousness ("Shadows")*. Professor Penrose's scientific training also made him wonder whether or not free will is just an illusion. As he stated in *Shadows*:

> *It might well be argued that each of our actions is ultimately determined by our inheritance and by our environment—or else by those numerous chance factors that continually affect*

---

46. Hawking (1988) 1996 ed., p. 71. Rather than saying we "know" the virtual particles exist, it would probably be more scientifically correct to say "we have evidence that the virtual particles exist."

our lives. Are not all of these influences "beyond our control," and therefore things for which we cannot ultimately be held responsible? . . .

If it is other than a mere convenience of language that we speak as though there were such an independent "self," then there must be an ingredient missing from our present-day physical understandings. The discovery of such an ingredient would surely profoundly alter our scientific outlook. . . .

I shall argue that when a "cause" is the effect of our conscious actions, then it must be something very subtle, certainly beyond computation, beyond chaos, and also beyond any purely random influences. Whether such a concept of "cause" could lead us any closer to an understanding of the profound issue (or the "illusion"?) of our free will is a matter for the future.[47]

There is relevance, also, to the question of "free will," as was raised at the end of section 1.11: might there be something that is beyond our inheritance, beyond environmental factors, and beyond chance influences—a separate "self" that has a profound role in controlling our actions? I believe that we are very far from an answer to this question. As far as the arguments of this book go, all that I could claim with any confidence would be that whatever is indeed involved must lie in principle beyond the capabilities of those devices that we presently call "computers."[48]

Again, we see in the above excerpts the intellectual struggle relative to the belief in free will that must occur if we first believe that humans are only made up of matter. Apparently Professor Penrose believes that "we are very far from an answer to this question" of whether or not humans have free will. Professor Penrose recognizes, however, that humans are capable of noncomputational thinking that cannot even "in principle" result from the

47. Penrose (1994), page 36.
48. Penrose (1994), page 401.

operation of a computer or anything that operates according to defined laws. As discussed in the chapter "Math," Professor Penrose is not able to conclude that humans have free will because he believes humans are only made up of matter.

Professor Penrose, however, also made statements in *Shadows* that would seem to imply that he does believe in free will. Note the following:

> . . . *if one chooses to abandon the methods of science at some point . . .*[49]

He apparently believes humans have free will. How else can people choose unless they have free will? Thus, Professor Penrose recognizes that humans have free will. Nevertheless, he is not willing to recognize that free will cannot be explained as a natural phenomenon that is subject to the rules of science.

Professor Penrose has a warning for those scientists who would try to develop a complete mathematical model of the universe that does not also explain human consciousness:

> *A scientific worldview which does not profoundly come to terms with the problem of conscious minds can have no serious pretensions of completeness.*[50]

I would comment that a scientific model of the universe which does not account for human consciousness *and* free will might be able to completely explain the *natural* world comprised of matter and energy but would not be able to completely explain human beings and their free will actions. It would also not be able to predict what happens in the *supernatural* and spiritual world. I have addressed Professor Penrose's warning by hypothesizing there is a soul-brain interface. Under this concept, the scientific model describes everything that happens in the natural world that is subject to the laws of science. Human free will would reside in the

---

49. Penrose (1994), page 145.
50. Penrose (1994), page 8.

supernatural world that is not subject to the laws of science. The soul interacts with the brain at the soul-brain interface to control the actions of the body and provide for the phenomena of human free will.[51]

CARL SAGAN AND ANDREA DURYA

Another popular scientist, Carl Sagan,[52] has also addressed the question of free will in the 1992 book, *Shadows of Our Ancestors ("Ancestors"),*[53] which he wrote jointly with his wife Andrea Durya who is also a scientist. One of the main themes of *Ancestors* is that human thought and actions are heavily influenced by our animal ancestors.

> *Might our penchant for imagining someone inside pulling the strings of the animal marionette be a peculiarly human way of viewing the world? Could our sense of executive control over ourselves, of pulling our own strings, be likewise illusory—at least most of the time, for most of what we do?*[54]

The above excerpt from *Ancestors* is another example of the intellectual conflict that must arise if we are to deny that humans have free will. The sentence seems to start out to claim that all of our actions are due to our natural bodily tendencies that we have presumably inherited from animal ancestors. But the authors interject the qualifiers "at least most of the time, for most of what we do" because they are not ready to completely give up on free will. Later in the book, the authors make a similar observation:

---

51. Potential mechanisms for this concept are described in the chapter "The Soul-Brain Interface."

52. Professor Sagan was host of "Cosmos," a popular 1980s television series about the universe. He was also the David Duncan professor of astronomy and space sciences and director of the Laboratory for Planetary Studies at Cornell University, distinguished visiting scientist at the Jet Propulsion Laboratory, California Institute of Technology, and the cofounder and president of the Planetary Society, the largest space-interest group in the world.

53. *Shadows of Our Ancestors* is discussed in more detail in the chapter "Biology."

54. Sagan (1992), page 168.

The fact that complex behavioral patterns can be trig-
gered by a tiny concentration of molecules coursing through
the bloodstream, and that different animals of the same spe-
cies generate different amounts of these hormones, is something
worth thinking about when it's time to judge such matters as
free will, individual responsibility, and law and order.[55]

Again, it appears that the authors might be making the claim that humans do not have free will, but it is not clear to me what their position is. I think most people agree that there are mitigating circumstances and influences that affect how our actions might be judged in a court of law. Attorneys, judges, and juries deal with such matters every day. Most people would agree that humans are influenced by hormones and emotions but that they are still free to decide how they will react to the feelings caused by the hormones and emotions. Judicial systems throughout the world are based on the assumption that humans do have free will and are responsible for their actions. In the final analysis, humans either do or do not have the capability to make free decisions whether or not they may be influenced by emotions and hormones and whether or not our bodies might do a lot of things automatically.

Even though a large percentage of the time our actions might be controlled by bodily functions or influenced by feelings, we are still free to accept or reject the impulses associated with those influences. For example, breathing is something that we do automatically without conscious choice because it is vital to our bodies staying alive. However, this does not keep us from choosing to stop breathing if we so wish. Our daily actions require a constant balancing between operating automatically on the one hand and freely choosing to do something on the other.

Other statements in *Ancestors* appear to indicate that the authors do believe humans have free will. Consider the following excerpt from *Ancestors*:

---

55. Sagan (1992), page 238.

> *We are free to posit, if we wish, that God is responsible for the laws of Nature, and that the divine will is worked through secondary causes. In biology those causes would have to include mutation and natural selection.*[56]

In order to be "free to posit, if we wish" we must have free will. Of course, I agree that humans do have free will, but it appears, as indicated by the following excerpt, that the authors of *Ancestors* are intent on making the case that humans are not significantly different from other animals even to the point of claiming that animals have free will and souls without any scientific evidence:

> *What, if anything, do the other animals think? What might they have to say if properly interrogated? When we examine some of them carefully, do we not find evidence of executive controls weighing alternatives, of branched contingency trees? When we consider the kinship of all life on Earth, is it plausible that humans have immortal souls and all other animals do not?*[57]

It is apparent that the authors are trying to make the case (without any real evidence) that whatever humans can do, it is no different in concept from what animals can do. If humans can think, so can animals. If humans can make free decisions, so can animals. If humans have souls, so do animals. We are left wondering, however, whether the authors are claiming that both humans and other animals have souls or that humans and other animals do not have souls. The authors, however, take note that there is no scientific basis for supernatural souls:

> *Almost all of [the great Western philosophers] believed that our distinction [from other animals] arises from something made neither of matter nor of energy that resides within the bodies of humans, but of no one else on Earth. No scientific evidence for such a "something" has ever been produced. Only a few of the great Western philosophers—David Hume, for*

---

56. Sagan (1992), page 63.
57. Sagan (1992), page 166.

instance—argued, as Darwin did, that the differences between our species and others were only of degree.[58]

So here we have the authors' position: there is no scientific basis for claiming that humans have supernatural souls. Whatever humans have, it is due to evolution and natural selection and is no different in concept from what the other animals have. It is different only in degree. In *Billions and Billions*, his last book before he died in 1996, Professor Sagan reaffirmed his belief that human life is explainable by physics and DNA.

> *The most significant aspect of the DNA story is that the fundamental processes of life now seem fully understandable in terms of physics and chemistry. No life force, no spirit, no soul seems to be involved. Likewise in neurophysiology: Tentatively, the mind seems to be the expression of the hundred trillion neural connections in the brain, plus a few simple chemicals.*[59]

I have the following comments:
- The scientific evidence for the "something" (a supernatural soul) referred to in the above excerpt from *Ancestors* is described in this book.
- In spite of the authors' claim of our physical kinship with animals, there is no scientific evidence that animals have either free will or supernatural souls.
- It is plausible that humans have supernatural souls and all other animals do not. If humans have supernatural souls, it is logical to conclude that it is something that has been given by an almighty supernatural being. Such an almighty being could choose to give supernatural souls to humans and not to other animals.
- It is possible for animals to make "decisions" that are completely determined by their biological makeup without having free will and without having supernatural souls.

---

58. Sagan, (1992), pg. 364.
59. Sagan, (1997), pg. 210.

- The authors of *Ancestors* did not provide an explanation as to how a human being could have a free will whose source is a natural phenomenon subject to the laws of chemistry and physics.

The above conclusions concerning human free will do not contradict the idea that animals are sentient creatures that can feel and sense their surroundings. The above conclusions do not contradict the idea that animals can feel pain and need to be protected from abusive treatment and unacceptable living conditions. If human beings do have free will, I believe this leads to the conclusion that they must be conscientious stewards of the environment and protect all plants and animals from improper treatment.

Francis Crick

In 1953, Francis Crick, a British biologist collaborated with American biologist James Watson on developing the theory concerning the shape of deoxyribonucleic acid ("DNA") and ribonucleic acid ("RNA"). In 1962, they shared the Nobel Prize for physiology along with Maurice Wilkins for their work on nucleic acid. DNA and RNA form the basis of the genetic code which transmits genetic information from one generation of plants and animals to the next generation. As described above in the excerpt from Professor Carl Sagan's book, many people believe DNA and RNA form the final scientific link needed to provide a complete material explanation for the existence and development of life. In his 1994 book, *The Astonishing Hypothesis: The Scientific Search for the Soul*, Dr. Crick confidently proclaims:

> *The Astonishing Hypothesis is that "You," your joys and your sorrows, your memories and your ambitions, your sense of personal identity and free will, are in fact no more than the behavior of a vast assembly of nerve cells and their associated molecules.*[60]

---

60. Crick (1994), pg. 3.

Dr. Crick's statement is similar to those described above. In his book, he explains his belief that human actions are determined by the "behavior of nerve cells and their associated molecules." In other words, free will must be an illusion because the interaction of molecules is subject to the laws of physics.

Dr. Crick's book contains "A Postscript on Free Will" that provides his explanation of free will. He describes free will in completely materialistic terms and describes the brain as a "machine" that "will appear to itself to have Free Will." He explains that the brain knows the reasons for some decisions and at other times the brain does not know or confabulates. Sometimes a small perturbation in the decision process will make a big difference in the end result, which "would give the appearance of the Will being 'free' since it would make the outcome essentially unpredictable." This is similar to the explanation of free will given by Professor Hawking as described earlier and by Professor Steven Weinberg as described in the chapter "Quantum Mechanics." However, *the outcome being unpredictable* is not a description of free will. Rather, true free will is the ability of a person to make a free choice among several options. Apparently, Professor Crick does not believe in true free will.

In my opinion, the most "astonishing" thing about Professor Crick's hypothesis is that a brilliant Nobel prize-winning scientist would abandon the idea that humans have free will. I cannot imagine that Dr. Crick lives his life believing that he has no choice over anything he thinks or does. Does he believe that the discovery of DNA was not due to his own careful evaluation of the scientific evidence and that it "just happened" with no true free will choices on his part? Was his writing of the subject book entirely devoid of any free will choices by him concerning what to study or what to say?

Most notably, Dr. Crick does not provide evidence that free will is an illusion. Rather his belief that free will is an illusion is based on a logical conclusion stemming from his *assumption* that humans are made up of only matter. I think he makes this assumption because he believes that believing in supernatural souls would be unscientific and akin to believing in myths.[61] He claims that most scientists, with the notable exception of Sir John Eccles (whose theories are discussed in the chapter "The Soul-Brain Interface"), do not believe in supernatural souls. However, as I have explained throughout this book, believing in supernatural souls is not unscientific.

Interestingly, Professor Crick spends one chapter of his book on optical illusions. He does this to show how difficult it is to describe neural processes in the brain. One of the optical illusions he covers is the Necker cube (reproduced in Figure 2.1 below) which is an illustration that includes the line edges of all sides of a simple cube, including the edges that we would not normally see if the cube was made of a solid, opaque material. There are two visual orientations of the Necker cube possible for the viewer to "see." These two orientations are also shown below by removing the edges that would not be visible if the cube were made of a solid, opaque material.

The cube is interesting to me relative to the question of free will in that the observer can look at the Necker cube with all of its edges visible and can *choose* to "see" in his or her mind either Orientation 1 or Orientation 2. This is free will in action.

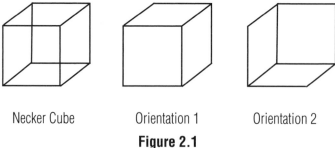

Necker Cube   Orientation 1   Orientation 2

Figure 2.1

61. Crick (1994), p. 4–6.

There are also several examples in Professor Crick's book wherein he speaks as though he believes people can actually make free choices. Some of these examples include:

"Some readers will find this approach disappointing since, as a matter of tactics, it deliberately leaves out many aspects of consciousness they would love to hear discussed . . ." (p. xi) For something to be deliberate, free will is required.

"I have tried to write for the general reader . . ." (p. xi) "Trying" requires free will.

"Don't be discouraged . . ." (p. xi) This phrase implies the reader has a choice whether or not to keep on trying if it is difficult to understand.

"I then state the assumptions I make (and the attitudes I take) . . ." (p. xii) "Attitude" requires free will.

"For readers wishing to follow up any topic . . ." (p. xiii) "Wishing" implies a free will choice.

"I shall be most grateful to readers who write to me to point out factual mistakes." (p. xiii) "Gratitude" is an attitude that is a free will choice. Some writers would be annoyed if people wrote to them pointing out mistakes. Dr. Crick has made a free will choice to be grateful.[62]

"Try for a moment to imagine this point of view." (p. 7) "Trying" requires a free will choice.

". . . many people are reluctant to accept . . ." (p. 7) "Reluctance" implies people have a choice whether or not to change what they believe in.

". . . one tries to select the most favorable system for the study of consciousness . . ." (p. 19) "Trying" requires a free will choice.

---

62. I hope he chooses to be grateful for my comments.

The preceding are just a few examples which indicate that Dr. Crick does believe that people can make free choices. In general, I would have to think that Dr. Crick believes in the scientific method which requires people to weigh the scientific evidence and then make a judgment and a free choice as to whether or not the evidence supports the hypothesis.

Dr. Crick also describes a case in which a woman who had an injury to the anterior cingulate sulcus part of the brain appeared to lose her will. Dr. Crick theorized that the anterior cingulate sulcus part of the brain is the source of free will. Since all parts of the brain are made up of matter, I would contend that no part of the brain can in itself be the source of free will. However, as described in the chapter "The Soul-Brain Interface," I believe it is logical to conclude there is a way that the soul interacts with the material brain to cause the soul's free will decisions to be implemented in the body by the central nervous system. Since the soul must interact with the brain, I would agree that the implementation of someone's free will could be impeded by damage to the brain. An analogy would be the inability to receive a radio or television signal because an antenna is damaged. The antenna is not the source of the radio or television signal. Likewise, damage to the brain in no way implies or provides evidence that the brain itself is the source of free will. The human brain is simply the mechanism by which free will decisions that originate in the soul are put into action.

Dr. Crick's book provides a good foundation for beginning to understand how the brain functions. It will be a valuable resource as we continue to assemble the pieces for understanding how the soul interfaces with the brain—one of the most exciting and important challenges of this millennium. Dr. Crick's book has a wonderful bibliography of books on consciousness, the mind-brain problem, and neural-physiology that anyone interested in pursing these topics should investigate. He provides a short description of each book along with his pithy comments.

MARVIN MINSKY

Marvin Minsky founded the Artificial Intelligence Laboratory at Massachusetts Institute of Technology where he is the Donner professor of science. He is a member of the National Academy of Science and a former president of the American Association for Artificial Intelligence. In his 1985 book *The Society of Mind*, he states that "[a]ccording to the modern scientific view, there is simply no room at all for 'freedom of the human will.'"[63]

By this he means that if you *assume* humans are made up only of matter, and if you *assume* anything that cannot be explained as a natural phenomenon does not exist, and if you *assume* humans do not have supernatural souls that are free to act unfettered by the laws of science and physics, then there is no room for human free will. It is not appropriate, however, to simply assume things for which there is no experimental evidence. On the contrary, evidence for the existence of human free will leads directly to the conclusion that humans do have supernatural souls.

Knowing the absurd world that results from a denial of free will, I find it incredible that a brilliant scientist such as Professor Minsky could so easily deny the existence of human free will. But, of course, he must deny free will if he first rejects the possibility of supernatural souls, believes that humans only have material bodies, and wants to remain intellectually honest.

From the perspective of artificial intelligence, it would be impossible to create a robot with free will. Thus, if humans do not have free will, it would be easier to make a robot with human-like characteristics.

CHAPTER CONCLUSIONS

1. Since all evidence to date indicates that natural phenomena are subject to scientific laws, a belief that humans are made up only of matter leads to the conclusion that humans do not have free will.

---

63. Minsky (1985), pg. 306.

2. Human free will cannot be explained as a natural phenomenon that is subject to the laws of nature. Human free will can be explained as a supernatural force that is not subject to scientific laws.

3. It is not practical to develop a mathematical model which can predict the actions of human beings. This, however, in no way implies that humans do not have free will.

4. The influence of natural bodily functions and feelings on human actions does not preclude the intervention of free will to allow someone to override the bodily feelings and to perform actions which oppose such feelings.

5. It is not scientifically justified to simply assume humans do not have free will.

6. It is not scientifically justified to simply assume humans do not have supernatural souls.

7. Apparent loss of free will and other mental faculties due to brain damage does not provide evidence that the brain is the source of free will and all mental faculties. It is possible for the free decisions and other mental faculties originating in a supernatural soul to be implemented through an interface between the soul and the brain. Brain damage would only indicate that the free will decisions originating in a supernatural soul are not able to be implemented via the brain.

ADDITIONAL READING ON FREE WILL[64]

Berofsky, Bernard, ed., *Free Will and Determinism* (New York and London, Harper & Row, 1966).

Brown, Geoffrey, *Minds, Brains, and Machines* (Bristol, Bristol Classical Press, 1989).

Dennett, Daniel C. *Elbow Room: the Varieties of Free Will Worth Wanting* (A Bradford Book The MIT Press Cambridge, MA, 1984).

Dworkin, Gerald, ed., *Determinism, Free Will, and Moral Responsibility* (Englewood Cliffs, N. J., and London, Prentice-Hall, 1970).

Fischer, John Martin, ed., *Moral Responsibility* (Ithaca, N.Y., and London, Cornell University Press, 1986).

Glover, Jonathan, *Responsibility* (London, Routledge & Kegan Paul, 1970).

Honderich, Ted, ed., *Essays on Freedom of Action* (London, Routledge & Kegan Paul, 1973).

Hook, Sidney, ed., *Determinism and Freedom in the Age of Modern Science* (New York, Collier, 1961).

Kenny, Anthony, *Free Will and Responsibility* (London, Routledge & Kegan Paul, 1978).

Morgenbesser, Sidney, and James Walsh, eds., *Free Will* (Englewood Cliffs, N.J., and London, Prentice-Hall, 1962).

Morris, Herbert, ed., *Freedom and Responsibility* (Stanford, Stanford University Press, 1961).

Sartre, Jean-Paul *Existentialism and Humanism* (London, Methuen, 1958).

Skinner, B.F., *Beyond Freedom and Dignity* (New York, Knopf, 1971).

Stevenson, Leslie, ed., *The Study of Human Nature* (New York and Oxford, Oxford University Press, 1981).

Watson, Gary, ed., *Free Will* (Oxford, Oxford University Press, 1982).

---

64. Additional reading list from Thornton (1989), pg. 138, partial list from *Bibliography of 20th Century Works.*

DILBERT reprinted by permission of United Feature Syndicate, Inc.

Chapter Three
FREE WILL TEST

In Chapter 1 of this book, I described why this world and human life would be absurd if humans did not have free will and all human actions and thoughts occurred only as the result of the laws of physics. Based on everyday experiences, most people would readily accept the premise that humans have free will. However, philosophers and spiritual thinkers have debated the existence and nature of free will for centuries. Also, as is evident from the excerpts in Chapter 2, "Free Will or Not," many of today's scientist might not accept the logic of philosophers and spiritual thinkers if such logic is not based on the scientific method. Thus, for the following reasons, it would be useful to have a test that would provide evidence for human free will:

- The hypothesis that humans have a supernatural soul, as presented in chapter 1, is based on humans having free will.
- The normal everyday actions of humans and the justice systems in all countries are based on the assumption that humans have free will and are able to make free choices.
- Some scientists have assumed humans are made up only of matter (which is subject to the laws of physics) and thus do not have free will.
- News articles appear from time to time purporting to show that some people are genetically predisposed for various behavioral afflictions such as alcoholism, drug

addiction, sex addiction, gambling addiction, violence, etc. In some cases, genetic predispositions are used to support the claim that humans cannot make free choices to overcome these afflictions and thus do not have free will.

Is it possible then to be able to provide evidence that humans have free will? Since the above important issues are affected by the existence (or absence) of free will, I have devised a test, as described below, to provide evidence that humans do have free will. Scientists use experiments to test scientific theories by making the experiments as identical as they need to be to show that the results of the test are reproducible. Likewise the test procedure described below is designed to control the circumstances surrounding the test such that the test taker faces identical circumstances going into Test Run 1 and Test Run 2 of the test. I will use two approaches to provide evidence that humans have free will:

- **Approach 1**—Approach 1 is based on the premise that, for multiple test runs, the outcome of each test run will be identical (within experimental accuracy) if the controlling circumstances for each test run are identical (within experimental accuracy) and if the test is performed on matter that is subject to the laws of physics. This is essentially the scientific method. Identical circumstances and identical inputs produce identical results. This occurs because atoms and molecules cannot decide on their own how they will react to their surroundings. Decades of scientific experiments provide evidence that all natural phenomena interact according to the rules of physics. On the other hand, if the controlling circumstances and inputs are identical for each test run but the results are diametrically different, this implies there is a force operating that is not subject to the laws of nature and that can decide on its own what to do. This ability to act unfettered by the laws of nature is what is known as "free will."

- **Approach 2**—Approach 2 is based on the premise that if we can identify all the possible explanations for the results of the test runs and then rule out all of the explanations except one, the one remaining explanation will be the best explanation available to date if there is no basis for discounting it. As described by physics professor Lawrence Krauss, this is the "Sherlock Holmes" method of investigation: "It is worth recalling Sherlock Holmes' adage that when you have eliminated all other possibilities, whatever remains, no matter how improbable, is the truth."[1] I agree with Professor Krauss' approach. However, note that although his statement might be appropriate for a detective such as Sherlock Holmes, it is not exactly appropriate for scientific investigation. A more correct scientific statement would be to say that whatever remains, no matter how improbable, is the best explanation available to date. Remember, as described in Chapter 1, that although each scientist (as well as every other human) has a belief system and uses science to seek the truth, science itself is not a belief system. Thus, it is not scientifically correct to say that an hypothesis has been "proved" and is the "truth."

As described below, I implemented a test and then attempted to explain the test results using explanations that would be consistent with a creature that is made only out of natural phenomena and does not have free will. However, none of the explanations that are based on a creature not having free will are consistent with the evidence. Of the explanations considered, only the explanation that is based on the test taker having free will is consistent with the evidence.

Note that since each person has a belief system, many people already believe that humans have free will. For such people, this

---

1. Krauss (1997), pg. 173.

chapter will provide additional evidence for their belief. Other people might not believe humans have free will or at least are reserving judgment on the matter until there is evidence for human free will. This chapter presents the sought-after evidence.

There is, of course, an inherent absurdity in trying to use evidence to convince people that humans have free will. Without free will, no one can freely decide whether or not the evidence supports the hypothesis that humans have free will. Without free will, scientists would not be able to freely decide whether or not any evidence supports any hypothesis. If I do not have a free will, then I did not have any real control over whether or not to write this book and which evidence to include. If people reading this book do not have free will, then they cannot freely decide whether or not they will change their minds and believe that people have free will. Without free will, they would not be able to freely decide whether or not there is sufficient evidence to believe in free will. Without free will, it would be absurd to say that someone has freely chosen to not believe in free will.

Summary of Free Will Test Procedure

The free will test uses the following test procedures:
- The test taker twice performs a set of actions as specified for the test under identical circumstances.
- The results of each test run are tabulated.
- As many explanations for the results as can be thought of are listed.
- Those potential explanations that are not supported by the evidence, are contradictory, or otherwise do not explain the results are discarded.

Under **Approach 1**, the fact that the results of the two test runs are not identical is evidence that the subject being tested (that is, a human being) has a "free will" that changes the brain and directs the material body to react differently under identical conditions.

Under **Approach 2**, all of the potential explanations that are based on natural phenomena without free will are shown to not be supported by evidence. This leaves free will as being the best explanation for the test results. To the extent we are confident that all the potential explanations for the test results were included in the original list and all explanations except one are shown to be not supported by the evidence, the explanation that is shown to be consistent with the evidence should be considered to be the best explanation available.

THE FREE WILL TEST

The following is a description of the free will test. Two pieces of paper are placed on a table. There is an "X" on one piece of paper and there is an "O" on the other piece of paper. It is explained to the test taker that the test is to see if she[2] will select one or the other of the two pieces of paper sixty times, if she will pick any other pattern she so chooses, or if she will pick no pattern at all. The test taker is not promised any reward or threatened with any punishment regardless of how she chooses.

At the beginning of each test run, the test taker marks down on a separate piece of paper how she will choose and hands the paper to the test giver. The test giver places the two pieces of paper on the table, one having an "X" and the other having an "O." The test taker chooses one of the pieces of paper and places it in a cup or basket. The test giver then checks which piece of paper is in the cup or basket, records which piece of paper was chosen, and returns the piece of paper to the table. The test giver puts the pieces of paper with the "X" and the "O" in different relative positions so that the position of the two pieces of paper does not affect the choice of the test taker. This is repeated sixty times for each test run.

---

2. To simplify the explanation, the test taker is referred to as female. However, several males also took the test.

ANALYSIS OF TEST RESULTS

Eight test takers performed the test. None of the test takers was aware of the test or the nature of the test until just before taking the test. The test results are summarized in the following table:

Table 3.1
Results of Free Will Tests[1]

	Test Run 1		Test Run 2	
	Choice Written on Paper at Beginning of Test Run	Pattern Actually Selected	Choice Written on Paper at Beginning of Test Run	Pattern Actually Selected
Test Taker 1	30 "O's" 30 "X's"	30 "O's" 30 "X's"	60 "O's"	60 "O's"
Test Taker 2	60 "X's"	60 "X's"	Alternate "X's" and "O's"	Alternate "X's" and "O's"
Test Taker 3	60 "O's"	60 "O's"	Random	Random order 33 "X's" & 27 "O's"
Test Taker 4	60 "X's"	60 "X's"	10 "O's" 10 "X's" (three times)	10 "O's" 10 "X's" (three times)
Test Taker 5	Alternate "X's" and "O's"	Alternate "X's" and "O's"	Alternate "X's" and "O's"	Alternate "X's" and "O's"
Test Taker 6	60 "X's"	60 "X's"	Alternate "X's" and "O's"	Alternate "X's" and "O's"
Test Taker 7	60 "X's"	60 "X's"	Refused to complete test[2]	
Test Taker 8	Alternate "O's" and "X's"	Alternate "O's" and "X's"	10 "X's" 10 "O's" (three times)	10 "X's" 10 "O's" (three times)

1. The tests were performed by members of St. Richard's Catholic Community on January 11, 2004; January 18, 2004; and January 25, 2004 in the early afternoon.

2. Although the test taker refused to perform Test Run 2, she and her spouse continued to sit and talk with the test giver for 10 to 15 minutes after the completion of Test Run 1.

In spite of the fact that conditions were identical for both test runs, the results of the two test runs were dramatically different for all test takers except Test Taker 5. Let us examine the potential ways we could explain how the Test Run 1 results and the Test Run 2 results could be dramatically different. The following are potential explanations of how one pattern could be followed during Test Run 1 and a completely different pattern could be followed during Test Run 2.

- *Alternative 1*— The test taker's "decision" to pick one way for Test Run 1 and another way for Test Run 2 was just an illusion. She does *not* have free will and cannot actually make a willful choice to do anything. She picked one way for Test Run 1 due to random chance and she picked another way for Test Run 2 due to random chance.

- *Alternative 2*—The test taker does *not* have free will, but she was programmed by another being to pick one way for Test Run 1 and another way for Test Run 2.

- *Alternative 3*—The test taker does *not* have free will, and she was forced by another being to pick one way for Test Run 1 and another way for Test Run 2.

- *Alternative 4*—The test taker does *not* have free will, but she was enticed by a reward (or threatened with a punishment) to pick one way for Test Run 1 and another way for Test Run 2.

- *Alternative 5*—The test taker does *not* have free will, but she was forced by evolutionary forces (natural selection) to pick one way for Test Run 1 and another way for Test Run 2.

- *Alternative 6*—The test taker does *not* have free will, but she was forced by environmental influences (nurture) to pick one way for Test Run 1 and another way for Test Run 2.

- *Alternative 7*—The test taker does *not* have free will, but her brain forced her to pick one way for Test Run

1 and another way for Test Run 2. However, the decision as to how to pick was not due to evolutionary forces or environmental influences.

- *Alternative 8*—The test taker *does* have free will, but she was enticed by a reward (or threatened with a punishment) or influenced by evolutionary forces (natural selection) or environmental factors to pick one way for Test Run 1 and another way for Test Run 2.
- *Alternative 9*—The test taker *does* have free will and freely chose to pick one way for Test Run 1 and another way for Test Run 2 without any outside influence.

ANALYSIS OF ALTERNATIVES

Alternative 1

Let us examine the first alternative which hypothesizes that the test taker does not have free will but randomly picked one way for Test Run 1 and randomly picked another way for Test Run 2. Consider first selecting 60 "X's" in a row. Intuitively we expect that there would be an extremely low probability that one of the pieces of paper could be randomly selected sixty times in a row. In fact, the likelihood of this occurring on a random basis is much less than you might imagine. Based on the application of statistical principals,[3] there is only one chance in a billion billion (a "1" followed by 18 zeroes) that the "X" piece of paper (for Test Run 1) would be selected all 60 times on a random basis. To understand how often a one in a billion billion event occurs, imagine there is a test program in which one piece of paper is selected randomly every five seconds and one test run (60 selections as described above) is completed once every five minutes. Imagine further that there is test run after test run, hour after hour, day after day, and year after year. Based on this "test" pro-

---

3. The application of statistical principles to this question is described in the appendix to this chapter.

gram, there would be one test run every 10 million million years (or 10,000 billion years) during which, on a random basis, the "X" piece of paper would be picked for all 60 selections. This period of time is 730 times longer than the currently accepted age of the universe which is estimated by some scientists to be about 15 billion years.

In other words, there is effectively a zero chance that the "X" piece of paper would be picked 60 times in a row due to random chance. (Try flipping a coin and see how long it takes to get either sixty heads or sixty tails in a row.) Because of the extremely low probability, this alternative is not a good explanation of how the "X" piece of paper could be picked 60 times in a row.

The statistical purist might make the comment that, if left to random selection, there is equal probability that any one of the billion billion combinations of 60 selections of "X's" and "O's" will occur and thus, 60 in a row is just as likely as any other combination. This is correct, but the important fact is that the test taker is able to predict at the beginning of the test that she will choose a certain pattern for the sixty selections. This ability to predict beforehand an occurrence that would have an extremely low probability if it were random indicates that it is *not* a random occurrence. The ability to predict at the beginning of the test any of the patterns that were chosen indicates that any pattern so chosen was not a random occurrence.

But that is not all. We can also imagine the test taker doing the same thing multiple times in a row. The test taker could follow Test Run 1 and Test Run 2 with more test runs of 60 choices in which she likewise chooses at the beginning of the test to pick a certain pattern and then does it. This could be repeated as often as you would like. Thus, we would see multiple consecutive occurrences, each of which would only be expected once every 10 million million years if they were occurring on a random basis. This is overwhelming evidence that the selection of a piece of paper 60 times according to a predicted pattern is not a random occurrence.

Each of the above free will tests could also be lengthened to require that more choices be made. Instead of picking a piece of paper 60 times, it could be lengthened to picking a piece of paper 100 times in a row, 1,000 times in a row or more. This would result in reducing even further the probability that choosing according to a predicted pattern is random.

Alternative 2

Next, we will examine Alternative 2, which postulates that the test taker does not have free will but was programmed by another being to pick one way for Test Run 1 and another way for Test Run 2. This alternative, of course, begs the question: *Who* programmed the test taker? If there is a programmer, that programmer must be someone with a free will who is choosing to affect the actions of someone else.

If this is the case, the test taker becomes nothing more than a biological computer. There is no evidence, however, in the methodology of the test as described above that would indicate the test taker was programmed by some other human. In addition, since this test has not been done before, it is not reasonable to think that someone else, unless he or she is an all-knowing spiritual being, would know that the test was going to be taken. Without knowing that the test was going to be taken, he or she would not have been able to program the test taker beforehand. Anyone reading this book could set up the test for himself or herself and take it, knowing that he or she was not programmed by someone else before the test.

We could theorize there is an almighty, all-knowing spiritual being who either controls human thoughts and actions moment to moment or has programmed all humans such that all human actions and thoughts have been predetermined. However, in my opinion, this is not a reasonable explanation. It does not seem reasonable that a spiritual being would create automatons, control their actions, and then give them the feeling that they make willful choices.

What would be the purpose for a spiritual being creating a universe and a race of creatures that are completely programmed or controlled? If such a spiritual being is all-knowing and all-powerful, He would know that He could make a complex human robot. If He programmed or controlled all the human robots, what would be the purpose? All human thoughts and actions would already be decided, and He would know what they would be. What would be His purpose in watching human robots move through a predetermined script?

Whether or not all human actions and thoughts have been programmed or controlled by a spiritual being, it does not seem likely that suddenly, after thousands of years of human existence, eight humans would be programmed to pick the same piece of paper 60 times in row or to pick the papers according to some specific pattern. Alternative 2 is similar to the centuries old debate as to whether and to what extent human actions are controlled or influenced by God. That debate will not be resolved here.

This book was written to show that free will choices cannot occur without a supernatural source. Since Alternative 2 presupposes a supernatural being, if you conclude Alternative 2 is the correct explanation, then you would likewise conclude there is a supernatural source for the observed actions.

Alternative 3

Alternative 3 is based on the hypothesis that the test taker does not have free will and has been forced to pick one way for Test Run 1 and another way for Test Run 2. "Forced" has two possible meanings:
- It could mean the test taker was threatened with severe physical consequences unless she picked the papers according to a required pattern. This meaning is not being considered under this alternative. It is considered under alternatives 4 and 8 below.
- Alternatively, "force" could mean that the individual is physically forced to do something. This would be

direct physical force applied to the person's body. An example would be when you take someone's hand and physically make that person move it in a certain way.

There is no evidence in the test procedures as described above that there was any physical exertion by some outside being to directly control the movements of the test taker. Once again, we could hypothesize that some unseen supernatural being used some unknown force to control the movements of the test taker, but this would be unreasonable for the same reasons as described under Alternative 2 above.

Alternative 4

Alternative 4 is that the test taker does *not* have free will but was enticed with a reward or threatened with a punishment to pick one way under Test Run 1 and another way under Test Run 2. This is not a logical or reasonable explanation because the test procedures do not include a reward or punishment that would explain the choice according to any pattern.

Typically, when an animal is trained to perform repetitive actions, there is some kind of biological reward that entices the animal to do something it would not otherwise have done. This reward can be in such forms as receiving food, being petted, or avoiding punishment such as use of a whip by a lion trainer. For the free will test described earlier there is no reward or punishment associated with picking any certain pattern. The test taker gets the same thing regardless of which pattern is chosen—nothing.

Alternative 5

Alternative 5 is that the test taker does *not* have free will but was influenced by evolutionary forces to pick one way for Test Run 1 and another way for Test Run 2. This is not a reasonable explanation because the results of taking the test do not in any way affect the ability of the test taker to survive and produce offspring. Natural selection (which is the basic mechanism of the theory of evolution) operates by genetically selecting over long periods of time the traits and characteristics that help an organism survive and produce offspring which will have the same traits or characteristics. For example, according to the theory of natural selection, over thousands of years, antelopes have developed a trait which allows them to recognize a stalking lion and then run away. This helps keep them alive, which enables them to produce offspring with the same trait. The weak and sick antelopes or those that through genetic mutation might not be able to recognize stalking lions would likely be caught and eaten by the lions. Thus they would not be able to have offspring and pass on any faulty traits.

For the test described above, there is no trait which would explain in evolutionary terms selecting one way for Test Run 1 and another way for Test Run 2. First of all, the test is not something that has developed over thousands of years. There is no evidence that it has ever been taken by anyone before now. Thus there is no possibility that taking the test could be some type of long-standing genetic ritual that affects the ability of the test taker to survive and produce offspring. Taking the test may require a certain level of intelligence which might have a genetic explanation. However the level of intelligence, in and of itself, does not explain the choice of any certain pattern for 60 selections. There might be evidence that genetic coding makes the test taker pick a certain pattern if the following conditions occurred:

- the test is taken regularly,
- for some unknown reason the person who by instinct picks according to a certain pattern is able to continue to live, and
- this happens time and time again.

But that has clearly not happened. Thus, the selection according to a certain pattern cannot be explained by evolutionary forces. In addition, if one pattern were to somehow provide a survival advantage under Test Run 1, then how would a completely different selection pattern for Test Run 2 be explained?

Alternative 6

Alternative 6 is that the test taker does *not* have free will but was influenced by environmental factors to pick one way for Test Run 1 and another way for Test Run 2. This is not a reasonable explanation because there is no evidence that anyone has ever been influenced by environmental factors to make 60 selections as described above. For this to be a valid explanation, we would expect to see people from time to time picking a piece of paper in the circumstances described above a number of times in a row or according to a certain pattern. Since this has not been observed, it is not reasonable to expect that someone would suddenly pick a piece of paper 60 times in a row (or according to some pattern) without some similar behavior occurring previously.

Alternative 7

Alternative 7 is that the test taker does *not* have free will, but her brain forced her to pick one way for Test Run 1 and another way for Test Run 2. However, the decision as to how to pick was not due to evolutionary forces or environmental influences. This is not a reasonable explanation because there are many aspects of the test that are not explained by this alternative.

Under this alternative, neither the test giver nor the test taker have free will, and the decisions to give the test and take the test are not due to evolutionary forces or environmental influences. Thus, there are no apparent reasons for either giving the test or taking the test, and both giving the test and taking the test "just happened." If the nature of the test were similar to other natural activities, it might be reasonable to conclude that they "just happened." However, the test is a very specific arrangement of actions: a piece of paper is chosen not five, not 10, not 21, not 28, not 33, not 46, not 52, and not any number between zero and 59 or more than 60, but is chosen exactly 60 times by multiple test takers. There is no apparent reason why anyone after the thousands of years of human civilization would suddenly give or take such a test. There is also no apparent reason why creatures without free will would "choose" to give or take such a test in which precisely 60 pieces of paper are chosen. Without any apparent reason for their occurrence, each step of the process adds another layer of improbability to these purposeless actions:

- the test giver asking and giving the test
- each test taker agreeing to take the test
- each test taker making a choice as to which pattern to pick
- each test taker beginning to pick according to that pattern
- each test taker continuing through exactly 60 picks to follow the pattern initially chosen
- all test takers except one choosing a different pattern for Test Run 2 (and one choosing not to complete the test)
- several of the test takers making a comment at the beginning of Test Run 2 such as "Let's see, what pattern shall I choose this time?"
- each test taker who completed Test Run 2 continuing to pick according to the pattern initially chosen for Test Run 2

The conditions present in the brain of each test taker that resulted in the pattern chosen for Test Run 1 would likewise be essentially the same conditions present for Test Run 2, but the results for Test Run 2 were dramatically different for all but one test taker.

The improbability of creatures without free will taking such a test and having alternatively the same and then dramatically different results for sequential test runs can be increased by adding multiple test runs (that is, Test Run 3, Test Run 4, etc.) to the process and/or increasing the number of pieces of paper chosen in each test run.

Alternative 8

Alternative 8 is that the test taker *does* have a free will, but she was enticed by a reward, threatened with a punishment, influenced by evolutionary forces (natural selection), or influenced by environmental factors to pick one way for Test Run 1 and another way for Test Run 2. For this alternative, no further explanation is really needed. It is already based on the premise that the human test taker has free will. The free will test is designed to determine whether or not humans have free will and not under what conditions or influences a person might be induced to pick certain pieces of paper. Thus, if the premise for this alternative is that humans have free will, it is already consistent with the main premise of this book: that humans do have free will. Nevertheless, as described above in the explanations for the previous alternatives, there is no evidence of a biological reward or threatened punishment. In addition, neither evolutionary nor environmental influences can explain the choice of a piece of paper 60 times according to a pattern.

Alternatives 1 through 8 Are Not Reasonable Explanations

The validity of this test is based on the following conclusions (and can be challenged if these conclusions are not accepted). I hope that the reader will agree:
- that the alternatives considered above include all potentially reasonable explanations for the study results and
- that Alternatives 1 through 8 are not reasonable explanations for the test taker taking the test and choosing one way for Test Run 1 and another way for Test Run 2.

This leaves us with only one viable alternative: Alternative 9.

Alternative 9

Alternative 9 is that the test taker has free will and freely chose to pick one way for Test Run 1 and another way for Test Run 2.

This is the only alternative that provides a reasonable explanation for the observed test results. Free will is the only reasonable explanation as to why someone would agree to take the test and go through the monotony and hassle of picking the same piece of paper 60 times (or more) in a row (or of picking some other pattern) without any expectation of reward. Free will is the only reasonable explanation as to why someone would pick one way for Test Run 1 and another way for Test Run 2.

CONDITIONS ARE NOT IDENTICAL

Another potential objection to the above test is that it is not a valid scientific test because the conditions at the start of Test Run 1 and Test Run 2 are not identical. This objection is based on the premise that the human brain is very complex and the conditions that affect it change from second to second. Since the inputs received by the brain and the status of the brain synapses change continuously, it is theorized that it is not possible to scientifically test human behavior or the human brain.[4] This objection is similar to Alternative 7 listed above, which hypothesizes that the results of the test are due to unexplained changes in the brain and are not due to free choices.

The answer to this objection is similar to the above comments about Alternative 7. The ability of the test takers to complete a test in exactly the manner described above is indicative that the brain does not change enough from second to second to make the test unscientific. If the brain changed as much as hypothesized for this objection, the test takers would not be able to complete the test as structured. The ability of test takers to make exactly 60 selections in agreement with the initial choice (as indicated at the beginning of each test run) is evidence that the brain does not change enough to make the test invalid. Several of the test takers even made selections that required keeping track of alternative

---

4. This, of course, would come as a shock to all of the thousands of scientists who do research on human behavior and the human brain.

quantities such as 10 in a row and 30 in a row. I do not doubt that alternative test procedures could be developed in which people choose other quantities or do multiple test runs with some choosing the same pattern for all test runs. I encourage others to develop and implement such other test procedures.

It is important to realize that free will can exist even if it is not possible to test it with a scientifically valid test. As indicated in Chapter 1, science is not a belief system. Ultimately, each person must decide whether or not the evidence for free will is sufficient to warrant a belief in free will (assuming of course that humans do have free will so that such a decision can be freely made). All of science is based on the premise that people do have the ability to freely choose whether or not the evidence supports a given hypothesis. Of course, if humans do not have free will, then nothing is done freely and I had no real choice whether or not to develop the test, the test takers did not have any real choice whether or not to take the test, and anyone reading this does not have any real choice whether or not to continue to read any more of this book or whether or not to believe in human free will.

DEFINITION OF FREE WILL

The above test provides evidence that humans do have free will. At this point, it makes sense to better define free will so that references to it are as unambiguous as possible. The following is a definition of free will based on scientific principles:

Free will is a power an organism has that enables it to be able to choose alternative actions under identical circumstances or even regardless of the circumstances.

The above definition demonstrates that free will is not subject to the laws of physics and is not a natural phenomenon because it violates decades of scientific evidence that for natural phenomena, given identical circumstances, identical forces, and identical initial conditions, the outcome will be identical. The free will test, on the other hand, allows the test taker to choose various combinations of pieces of paper regardless of the conditions under which the choice is made. The test taker can choose:

- all "X's"
- all "O's"
- 10 "X's" and then 10 "O's" and then 10 "X's" again or
- any combination she so "chooses."

There is no meaningful distinction between the conditions that exist when 60 "X's" are chosen and a few minutes later when 60 "O's" or another combination are chosen. The conditions are identical before the selection of each combination of papers. Even if there are any theoretical differences in the conditions under which the choices are made, they do not explain why one time 60 "X's" are chosen and then 60 "O's" (or some other pattern) are chosen.

Also, if the claim is that there will always be a theoretical difference between two sets of conditions, this would preclude scientists from ever performing experiments to test scientific theories. In general, conditions should be as identical as they need to be so that the aspects being tested by the experiment are not affected.

ARTIFICIAL INTELLIGENCE

A question comes to mind concerning the free will test: can the free will test be used to determine if the test taker is human? In the 1930s, Alan Turing, a mathematician, first conceptualized the idea of a "universal computing device." These "Turing machines" have since been developed and they are commonly called "computers." One long-standing puzzle that has been posed is whether there is any test or any group of questions that could be used to determine whether you are communicating with a human or a Turing machine (that is, a computer). This is known as the Turing test and is based on a similar challenge posed by the philosopher René Descartes in the 1600s.

The concept that an advanced computer or robot could someday be human-like has been popularized in many science fiction stories. A few of the more famous examples include the android named C3PO in the 1979 film *Star Wars*, the Terminator in the 1984 movie by the same name, and the humanoid named Andrew in the 1999 film *Bicentennial Man*. As machines with "artificial intelligence" or "AI" become more and more complex and sophisticated, some people have theorized that it will eventually be useful if there would be a way to determine whether a conversation is with a human or a computer/robot.

I am sure some people in the AI community will read about the above free will test and will remark that there will someday be a robot that can pick the "X" piece of paper 60 or 2,000 times in a row in precisely the way described above. This robot will look and act just like a human. Does that mean the robot has a free will and a soul? Is there a way to use the free will test to determine whether or not the test taker is human? In response to these questions I would offer the following observations:

- If a robot "chooses" to pick the same piece of paper 2,000 times in a row, it is not because the robot has a free will. It will be because the robot has been programmed to pick the same piece of paper 2,000 times in a row or will be following the direction of its computer brain to pick the same piece of paper 2,000 times in a row. The robot might even have "learned" to pick papers in such a situation based on similar experiences and a complex algorithm or computer program that controls the robot.

- The actions of the robot would beg the question: who programmed or designed the computer brain of the robot? Whoever programmed or designed the computer brain of the robot has a free will to be able to choose to influence the outcome of a test.

- However, from the outside looking in, we might not be able to tell whether or not we are looking at a human taking the test or a sophisticated, complex robot.

The free will test works on humans because there is no evidence that humans have been programmed to pick the same piece of paper 60 or 2,000 times in a row. Using the free will test to provide evidence that humans have free will requires, of course, that we know the test taker is human. If the test taker is not human, we do not know whether or not the test taker has been programmed or trained to pick the same piece of paper multiple times in a row or is following the dictates of its computer brain. Thus, the free will test is not a solution to the Turing machine puzzle. The free will test is used to provide evidence that humans have free will and not to determine whether or not a test taker is a human or a robot. Since a robot acts according to how it is programmed and according to the dictates of its computer brain, by definition it cannot exercise free will. The addition of a random number generator would make it possible for the robot to make a decision without either the robot, itself, or an outsider knowing what the robot will do in a certain situation. A random action, however, is not the same as a free action.

Likewise, an animal might be trained to pick one item multiple times in a row, but that would not be an example of free will because the animal would be reacting to the expectation of a reward or the threat of punishment.

The free will test cannot be used to determine whether or not a human being with a mental or brain deficiency (either due to genetics, disease, or an injury) has free will. If a person cannot take the free will test or does not understand the test, there are no test results and the test is not conclusive. If a person cannot take the test, it only means that there is a reason the person cannot take the test. It does not indicate whether or not the person has free will. As an analogy, to an observer, a sleeping person does not appear to make free choices. This, however, does not mean the sleeping person lacks the capability to make free decisions.

Many Worlds Objection

There is a radical interpretation of quantum mechanics[5] which is known as the "many worlds" theory of quantum mechanics. Under the many worlds theory, there are an infinite number of locations for each atomic particle at any moment in time but each location has higher or lower probability of occurrence based on the quantum wave equation. According to this theory, all possibilities happen (with varying degrees of probability), which results in the creation of an infinite number of parallel worlds. Thus, for example, the test taker not only picked 60 "X's" in a row, she also picked every other combination in other parallel worlds. Thus, someone might claim that the free will test does not provide evidence of free will because all other possibilities also happened. In response to this, I would make the following observations:

- The evidence from the tests taken indicates that the people taking the tests performed as they predicted. There is no evidence of "other worlds" in which they did not perform as they predicted or in which they chose all of the other billion billion possible patterns.

- Assuming that there are "other worlds" does not discount the evidence that the experimenter in this world made a choice to pick 60 "X's" in a row and documented it before actually doing it. This is evidence that an actual free choice has been made. Otherwise, why should the documented choice at the beginning of the experiment match the outcome? If the "other worlds" happen probabilistically, there would be no reason for the documented choice at the beginning of the test to match the actual pieces of paper picked. The fact that the documented choice before the test consistently matches the actual pieces of paper picked is evidence of a free choice. If there are "other worlds,"

_____

5. For a discussion of the principles of quantum mechanics, see the chapter "Quantum Mechanics."

they are sterile worlds because no one is aware of them and no one affects them by free choices.

FREE WILL MUST BE FREE

It is logical to conclude that the very nature of free will precludes it from being subject to any present or future laws of science. If free will were subject to any laws of science, it would not be "free" and humans would not be able to choose alternatives as described in the preceding free will test.

Additional scientific evidence for human free will can be found in the chapter "The Soul-Brain Interface" in the discussion on obsessive-compulsive disorders.

CHAPTER CONCLUSIONS

1. If a human test taker declares that she will pick the same piece of paper 60 times in a row (or will select some other pattern) and then does it, it is not a reasonable explanation that a certain pattern is chosen due to random chance.

2. There is no evidence that the human test takers in the free will test were programmed or forced by another human being or supernatural being to pick the same piece of paper 60 times in a row (or to pick according to some other pattern). Thus, it is not a reasonable explanation that the same piece of paper was chosen 60 times in a row (or that the papers were picked according to some other pattern) because the human test takers were programmed or forced by another human being or a supernatural being to do so.

3. If the human test takers were programmed or forced by another human being or a supernatural being, such being would have to have free will.

4. The free will test was designed to prevent the test takers from being enticed by a reward or threatened

by a punishment. Thus, it is not a reasonable explanation that the human test takers picked the same piece of paper 60 times in a row (or picked the papers according to some other pattern) by being enticed by a reward or threatened by a punishment.

5. Choosing the same piece of paper 60 times in a row (or picking according to some other pattern) does not confer any survival benefits on the test takers. Thus, it is not a reasonable explanation that the human test takers were influenced by evolutionary forces to choose the same piece of paper 60 times in a row (or to pick according to some other pattern).

6. There is no evidence that anyone previously in the history of human civilization has performed the free will test described in this chapter and picked the same piece of paper any number of times in a row or according to any pattern. Thus there is no evidence that any of the test takers were influenced by environmental factors (nurture) to pick the same piece of paper any number of times in a row or according to any pattern.

7. It is not a reasonable explanation that it "just happened" for no apparent reason that the test takers picked the same piece of paper 60 times in a row (or according to some other pattern) and then picked the papers 60 times according to another pattern.

8. The only reasonable explanation for someone picking the same piece of paper 60 times in a row (or picking according to some pattern) and then picking the papers 60 times according to another pattern is that the human test taker has a free will and freely chose to follow the pattern.

9. The free will test requires that we know the test taker is a human being.

10. The free will test cannot be used to determine whether or not the test taker is a human.

11. The inability of a person to perform the free will test does not indicate that the person does not have free will.

Appendix to Free Will Test Chapter

Explanation of Probability Theory

The easiest way to understand probability is to look at a simple system and then extrapolate it to a larger situation. Let us consider the classic example of flipping a two-sided coin with one side "heads" and the other side "tails." For each flip, there is an equal probability of the coin landing heads ("H") or tails ("T"), that is, there is a 50 percent probability of heads and a 50 percent probability of tails. We have the following possibilities for one flip:

> **Possibility 1:** H
> **Possibility 2:** T

The number of possibilities for one flip is equal to 2. In mathematical terms, this is 2 raised to the power of 1, which is designated as 2^1. If we flip the coin twice we have the following possibilities:

> **Possibility 1:** H H
> **Possibility 2:** H T
> **Possibility 3:** T T
> **Possibility 4:** T H

It should be apparent that these are the only four possible combinations and that each combination has the same likelihood, that is, the same probability of happening. Each combination has a one out of four or a 25 percent probability of happening. Thus if we flip the coin twice, there is a 25 percent probability that it will be HH, a 25 percent probability that it will be HT, a 25 per-cent probability that it will be TT, and a 25 percent probability that it will be TH. Let us assume we pick one side of the coin (say, for example, heads) and then seek to determine the probability that the one side we pick will occur for every coin toss. As shown above, if we flip the coin two times we have 2 x 2 or 4 possible combinations. This is 2 raised to the power of 2, which is designated 2^2. Thus, for two coin tosses, there is a 25 percent probability that heads will occur for both coin tosses, (that is, HH). This means that if we perform a test which includes two flips of a coin and repeat the two-flip test 100 times, we would expect to have close to 25 of the tests be HH. Likewise, we would expect to have close to 25 of the tests be HT, close to 25 of the tests be TT, and close to 25 of the tests be TH. As we increase the number of two-flip tests, the percentages in each category should get closer and closer to 25 percent. For example, consider the following typical results of a two-flip test:

Table 3.2
Typical Results of Two-Flip Tests

	100 Two Flip Tests		1,000,000 Two-Flip Tests	
	Quantity	Percent of Total	Quantity	Percent of Total
Number of Tests	100		1,000,000	
Number of Flips	200		2,000,000	
Number of HH	25	25%	250,420	25.04%
Number of HT	26	26%	250,682	25.07%
Number of TT	20	20%	249,635	24.96%
Number of TH	29	29%	249,263	24.93%
Total Number of H's	105	52.5%	1,000,785	50.04%
Total Number of T's	95	47.5%	999,215	49.96%

As shown in the above table, each combination of two flips occurs about 25 percent of the time. For the example of 100 tests (first two columns) the number of HH and HT sequences, is only a few more than the TT and TH sequences and the associated percentage of occurrence for each combination is several percentage points above or below 25 percent. As the number of tests increases to 1,000,000 (last two columns) the percentage of occurrence of each combination is only a few hundredths of a percent above or below 25 percent even though there are several hundred more HH and HT sequences than TT and TH sequences. As the number of tests increases even higher, the percentage occurrence of each sequence would be expected to get even closer to 25 percent.

Note that the previous table indicates that it is not possible to know what the outcome of any single two-flip test will be. If each flip occurs on a random basis, only the probability of the sequence will be known. Thus if before a two-flip test the flipper predicts that it will be two heads in a row, there would be a 25 percent probability of getting two heads in a row. There would likewise be the same 25 percent probability of getting one of the other three combinations (HT, TT, or TH). If HH is predicted and HH occurs, we are not impressed by the predictive power of the flipper because a 25 percent probability is reasonably high enough that it could likely occur. As the number of flips in the test increases and the associated probability of any one sequence decreases, the more surprised we become if the prediction before the test matches the outcome.

Look again, however, at the four possible results of flipping a coin two times that are listed above. In these four possibilities, there are four heads and four tails. This means there is an equal probability of getting heads and tails. Assume we perform the two-flip test many times and count the total number of heads and tails, regardless of the sequence in which they occur. As we perform more and more two-flip tests, the total number of heads and tails will get closer and closer to 50 percent and 50 percent. This is likewise indicated in the previous table entitled "Typical Results of Two-Flip Tests."

Thus it is important to clarify which probability we are seeking as we start the test:
- There is a 25 percent probability of getting two heads in a row.

- There is a 25 percent probability of getting two tails in a row.
- There is a 50 percent probability of getting heads if the sequence is not considered.
- There is a 50 percent probability of getting tails if the sequence is not considered.

If we flip the coin three times, the possibilities go up geometrically to an amount equal to 2 x 2 x 2 or 8. This is 2 raised to the power of 3, which is designated as 2^3. The possible outcomes are listed as follows:

Possibility 1: H H H

Possibility 2: H H T

Possibility 3: H T H

Possibility 4: H T T

Possibility 5: T T T

Possibility 6: T T H

Possibility 7: T H T

Possibility 8: T H H

If we toss the coin three times, there is a one in eight probability (that is, a 12.5 percent probability) that heads will occur every time (that is, HHH). This trend continues so that the number of possible combinations is equal to 2 raised to the power equal to the number of flips. Thus, for 4 flips of the coin we have 16 possible combinations (this is 2 raised to the power of 4 or 2^4), for 5 flips of the coin we have 32 possible combinations (this is 2 raised to the power of 5 or 2^5), etc. Likewise, for 4 flips, we have a one in 16 probability that heads will occur every time (that is, HHHH). A one in 16 probability is equal to 6.25 percent. For 5 flips we have a one in 32 probability (that is, a 3.125 poercent probability) that heads will occur for all flips (that is, HHHHH).

We can see that as the number of flips increases, the probability that heads will occur for all flips decreases quickly. The following table summarizes the probability of getting the same side of the coin for every flip as the number of flips increases.

Table 3.3
Probability Associated with Multiple Flips of a Coin

Number of Coin Flips	Chance of Getting All Heads = One Chance in:	Probability of Getting All Heads
1	2	50%
2	4	25%
3	8	12.5%
4	16	6.25%
5	32	3.125%
6	64	1.56%
7	128	0.78%
8	256	0.39%
9	512	0.20%
10	1,024	0.10%
20	1,000,000	0.0001%
30	1,100,000,000	0.0000001%
40	1,100,000,000,000	0.0000000001%
50	1,100,000,000,000,000	0.0000000000001%
60	1,200,000,000,000,000,000	0.0000000000000001%

As shown in the above table, by the time we get to 30 flips, the probability that heads will occur for each and every flip is one chance in about 1 billion tries. Thus, if flipping the coin 30 times is considered as one "try," there would be—on average—one try in which all flips produce 30 heads for every 1 billion tries. If there are 60 flips in one try, there would be—on average—one try in which all flips produce 60 heads for every 1 billion billion tries.

When there are two pieces of paper to choose from, as in the free will test described in this chapter, picking one piece of paper by chance has the same probability as flipping a coin, (that is, 50 percent). Thus the above examples which use "heads" and "tails" would likewise be applicable to "X" and "O" pieces of paper if they are selected on a random basis. Thus, if the pieces of paper are chosen on a purely random basis, there would be one time in 1 billion billion tries that all 60 pieces of paper are "X" and one time in 1 billion billion tries that all 60 pieces of paper are "O." If the pieces are chosen randomly, each combination of 60 "X's" and "O's" would occur on average one time in every 1 billion billion tries.

"There exists no rational evidence of biological mechanisms that can produce 'free will' in humans."

—William Provine, professor of the history of biology, Cornell University

"I verily believe, free will and chance are synonymous."

—Charles Darwin

In the United States, many Christians put the following "FISH" symbol on their cars to indicate that they are Christian:

The FISH symbol is so used because it is believed that early Greek-speaking Christians used the FISH symbol to indicate they were Christians. They used this symbol because the Greek word for "fish" is "icthus" which is an acronym for Jesus Christ, Son of God, Our Savior as follows: ΙΧθΥΣ (Greek letters) = ICHTHUS (transliteration) = ancient Greek word for fish.

I	=	J	=	Jesus		
X	=	CH	=	Christ		
θ	=	TH	=	Theou	=	of God
Y	=	U	=	(H)uius	=	Son
Σ	=	S	=	Soter	=	Savior

Because of the theory that life originated in the sea and evolved to land animals (and possibly from land animals back to the sea), many believers of the theory of evolution have co-opted the FISH (ICHTHUS) symbol to express their faith in the theory of Darwinian evolution as follows:

Chapter Four
BIOLOGY

Biology is the study of life, including human life. Science is a discipline that seeks to discover evidence which can be used to form conclusions as to the rules that govern the interaction of natural phenomena. The science of biology seeks to discover the rules that govern all aspects of the existence and functions of living organisms.

The bodies of all animals (including humans and human brains) are made up of atoms and molecules. All of the biological mechanisms of animals function using the atoms and molecules that make up their bodies. Thus, all biological mechanisms (including the functioning of the human brain) are subject to the laws of chemistry and physics which govern the interaction of atoms and molecules. As described in more detail in the chapter "Quantum Mechanics," the interaction of atoms and molecules according to the laws of chemistry and physics does not allow any room for free will. In order for free will to allow for truly free actions, it must not be subject to any rules. This leads directly to the conclusion that human free will cannot be explained as a biological mechanism.

Cornell University Professor William Provine,[1] a leading historian of science agrees that human free will cannot be explained

---

1. William B. Provine is professor of the history of biology in the section of ecology and systematics, division of biological sciences, and in the department of history at Cornell University.

as a biological mechanism. As he stated in a 1993 article in the journal *Biology and Philosophy*:

> *There exists no rational evidence of biological mechanisms that can produce "free will" in humans.*[2]

I admire the intellectual honesty of Professor Provine in stating that free will cannot be explained as a biological mechanism. I am also thankful because his statement helps confirm the previously asserted understanding as to why free will cannot be explained as a biological mechanism. Of course, if humans have free will and free will cannot be explained as a natural biological mechanism, this leads to the logical conclusion that the source of free will must be a supernatural phenomenon. However, as discussed below, Professor Provine also rejects the existence of supernatural reality. This means that for Professor Provine there is neither a natural nor a supernatural source for free will. Thus, his beliefs imply that human free will is just an illusion. I would caution: if humans do not have free will, everything quickly becomes absurd. Without free will, how can you even choose what you will or will not believe? How can you choose whether or not to believe in free will?

Professor Provine's beliefs (which have apparently been forced on him because he does not have the free will to choose otherwise)[3] are based on a complete rejection of the supernatural world:

> *First, modern science directly implies that the world is organized strictly in accordance with mechanistic principles. There are no purposive principles whatsoever in nature. There are no gods and no designing forces that are rationally detectable. . . .*

---

2. *Biology and Philosophy* 8, 123, 1993. The text of the complete article can be found at *http://www.vuletic.com/hume/provine.html*

3. Don't blame me for making sarcastic remarks. My heredity interacting with environmental influences caused me to write every word in this book. But if you do blame me, how can it be otherwise because your heredity and environmental influences cause you to think what you think?

Second, modern science directly implies that there are no inherent moral or ethical laws, no absolute guiding principles for human society.

Third, human beings are marvelously complex machines. The individual human becomes an ethical person by means of two primary mechanisms: heredity interacting with environmental influences. That is all there is. . . .

Fourth, we must conclude that when we die, we die and that is the end of us. That is what modern science tells us.[4]

As I have commented elsewhere in this book, I think that Professor Provine has misinterpreted the nature of science. Science only determines the laws which govern the natural world. Science does not claim there is no supernatural world, nor does it seek to describe the characteristics of the supernatural world. Science does *not* tell us that "when we die, we die and that is the end of us." Also, I call on all scientists to consider how the world quickly becomes absurd and meaningless if humans do not have free will. Without free will, we would not be free to decide whether or not to pursue scientific investigation and whether or not to believe in any specific results of scientific investigation. Professor Provine believes humans are shaped by their heredity (nature) and environmental influences (nurture) and "that is all there is." The implication of Professor Provine's beliefs is that becoming an ethical person is outside of our control. How then can we be guilty of any crime? How can we choose to be a moral person? Note also that the evidence for human free will[5] leads to the conclusion that there is a supernatural entity that is the source of such free will. Such evidence is not consistent with Professor Provine's *assumptions* that human free will is an illusion and that there is no supernatural reality.

---

4. Goldman (1989), pg. 261–262, included in William Provine's essay "Evolution and the Foundation of Ethics."

5. See the chapter "Free Will Test."

Professor Provine's belief that free will cannot be explained as a biological mechanism puts him in good biological company. Charles Darwin also believed that free will cannot be explained as a biological mechanism. As described by Robert J. Richards in his 1987 book *Darwin and the Emergence of Evolutionary Theories of Mind and Behavior*, Mr. Darwin equated free will with chance:

> *Darwin realized that a biological explanation of thought and behavior implied that organisms acted under law, that they were not free. Free will could only be equivalent to chance: "I verily believe," he remarked in his M Notebook, "free will & chance are synonymous. Shake ten thousand grains of sand together & one will be uppermost, so in thoughts, one will rise according to law."*[6]

As described in the chapter "Free Will Test," chance occurrences are not the same as free, deliberate actions. I think that Mr. Darwin based his belief that true free will does not exist on an *assumption* that only natural phenomena exist.

Emerging Properties

Some philosophers and scientists hypothesize that the conscious mind has emerged as a natural biological phenomenon through evolution by natural selection. Under this theory, as animals became more aware and more conscious of their surrounding environment, they were better able to react to conditions that threaten their survival. Thus, by becoming more conscious, they increased their survival rate and the genes for consciousness tended to increase in the population. However, even if "consciousness" has emerged as a natural biological property of the higher animals, free will, as discussed above, cannot be explained as a biological mechanism. This is because all biological mechanisms function under the laws of chemistry and physics, which

---

6. Richards (1987), pg. 122, quoting from page 31 of Charles Darwin's *M Notebook* which was transcribed by Paul Barrett and included in the 1974 book *Darwin on Man* by Howard Gruber on page 271.

leaves no room for free will. The effect this conclusion has on the theory of evolution by natural selection is discussed below.

It is important to recognize that the word "consciousness" has many meanings. "Consciousness" not only means being aware of the surrounding environment but can also mean being aware of being aware. This has been referred to as the "Cartesian Theater" wherein people can step back, as it were, and view themselves in the situation. I think this ability to "step back" and consider alternatives is an important part of making free decisions. Based on this, it is doubtful to me that this "being aware of being aware" aspect of consciousness is an emergent property. I think it is important to be aware that being aware of being aware is not a natural phenomenon.

EVOLUTION BY NATURAL SELECTION

Very few scientific theories have evoked such passion and controversy among the general public as the theory of evolution by natural selection.[7] Since it was first presented for general consumption by Charles Darwin in his 1859 book *On the Origin of Species by Means of Natural Selection,*[8] there have been passionate voices both for and against the theory. It caught the attention of the world in the 1925 "monkey" trial in Tennessee during which John Scopes, a high school teacher, was charged with violating a state law that prohibited the teaching of evolution.[9] This con-

---

7. Throughout this book, the term "evolution" refers to the process of the modification of organisms that occurs over long periods of time through genetic variation and genetic mutation and the subsequent selection of beneficial traits by natural selection. There had been other theories of evolution proposed by other scientists before Mr. Darwin, but his innovation was the mechanism of natural selection. The concepts of genetic variation and genetic mutation were developed after Mr. Darwin.

8. The full title is *On the Origin of Species by Means of Natural Selection or the Preservation of Favoured Races in the Struggle of Life.* The full text of the first edition of the book can be found at *http://pages.britishlibrary.net/charles.darwin/texts/origin1859/origin07.html*

9. Mr. Scopes was found guilty and fined $100, but the conviction was overturned on a technicality. The Tennessee law against teaching evolution remained "on the books" until 1967 when it was rescinded by the state Legislature.

flict was dramatized in the 1960 movie *Inherit the Wind* starring Spencer Tracy as the character representing Clarence Darrow,[10] the attorney defending the teacher, and Frederic March as the character representing William Jennings Bryan, the prosecuting attorney. It is an interesting twist that the underlying reason Mr. Scopes was charged with a crime is reminiscent of the charge against Socrates in ancient Greece: corrupting the youth. The dichotomy, however, is striking. Mr. Scopes was teaching a theory that many felt would lead to the abandonment of spiritual principles while Socrates thought he was teaching others how to be more spiritual.

By the time of Mr. Darwin's death in 1882, most of the scientific community had accepted evolution (by some mechanism) as an explanation for the origin of species. By the mid-1900s, with the synthesis of genetics and evolutionary biology, most of the scientific community had come to recognize that the theory of evolution by natural selection can explain a large body of scientific evidence. A large percentage of people, both in the scientific community and general population had come to believe that the evolution of species, as described by Mr. Darwin, had occurred. Beginning in the late 1900s, the controversy has been rekindled by the claims by some scientists and others that there is evidence which is not consistent with the theory of evolution. This has led to the demand in some areas of the United States that whenever evolution is taught in public schools, other evidence which is pur-

---

10. In 1924, Clarence Darrow had also been the defense attorney for the notorious case of Nathan Leopold and Richard Loeb. These two intelligent, well-to-do teenagers had kidnapped and murdered a child. In his closing arguments, Mr. Darrow implied that the two boys were not actually guilty because their actions were due to their genetics (nature) and their environmental influences (nurture). In a statement similar to that of Professor William Provine quoted above in this chapter, Mr. Darrow criticized the people "who seriously say that for what Nature has done, for what life has done, for what training has done, you should hang these boys." For a summary of Mr. Darrow's closing statement, see *http://www.law.umkc.edu/faculty/projects/ftrials/leoploeb/leopold.htm*

ported to discredit evolution be taught as well. Such evidence is claimed to support various theories such as:

- The theory that the age of the Earth is only several thousand years old and that all biological species came into existence at the same time (the theory of special creation which some refer to as "creation science"), or

- The theory that certain biological features are too complex to have evolved and needed to have been created by an intelligent being (the theory of intelligent design).

The theory of special creation and the theory of intelligent design are described in more detail later in this chapter. A number of books have been written in recent years that present the evidence the authors claim discredits the theory of evolution.[11] There has been ongoing controversy over what constitutes valid scientific evidence, who decides what is valid evidence, and whether or not state Legislatures should pass laws that mandate public school students be exposed to certain evidence and theories.

Why does the theory of evolution evoke such emotions? Most likely the theory's claims about the origin of humans and the resulting implications concerning the place of humans in the world and in the cosmos touch a sensitive spot in the human psyche. In the simplest terms, many who believe that the theory of evolution is correct have reached the conclusion that humans are only the product of natural causes and are not intrinsically different from any other species. Based on the theory of evolu-

---

11. See for example, *Of Pandas and People: The Central Question of Biological Origins* by Percival Davis and Dean Kenyon (1989); *Darwin on Trial* by Phillip Johnson (1991), *The Creation Hypothesis*, edited by J. P. Moreland (1994); *Darwin's Black Box: The Biochemical Challenge to Evolution* by Michael Behe (1996); *Mere Creation: Science, Faith, and Intelligent Design* by Willian Dembski (1998); *Icons of Evolution: Science or Myth?: Why Much of What We Teach about Evolution Is Wrong* by Johnathan Wells (2000); and *Intelligent Design Creationism and Its Critics: Philosophical, Theological, and Scientific Perspectives*, Robert T. Pennock, ed. (2001).

tion, humans and other primates[12] (such as monkeys and apes) evolved from a common ancestor that lived millions of years ago. As described below, this would make humans and the other primates distant relatives.

Occasionally, people erroneously describe evolution as the theory that "humans evolved from apes." This is not accurate. Rather, the theory of evolution teaches that humans and apes are different branches which developed in parallel from a common ancestor. That is why both humans and apes still exist today.

People who have a common ancestor are cousins. For example, people with the same grandparents are called first cousins, people with the same great grandparents are second cousins, people with the same great-great grandparents are third cousins, etc. If we were to assume the first human couple lived 200,000 years ago[13] and each generation was on average 25 years long, the current generation would be the 8,000th generation of humans. This would imply that any two people on earth are cousins, ranging from first cousins to approximately 8,000th cousins. Similarly, if humans and other primates have a common primate ancestor that lived 10,000,000 years ago, with average generations of 25 years, the current generation of humans and other primates would be the 400,000th generation of primates. All humans would thus be (more or less) 400,000th cousins of other primates on earth. Many people are not comfortable calling monkeys and chimpanzees their "cousins" no matter how remote the relationship.

The theory of evolution also claims that going back further in time, humans and all mammals have a common ancestor. If the theory is correct, the legendary storyteller Uncle Amos may have

---

12. Primates are the most highly developed order of mammals including human beings and other mammals that closely resemble humans. They include about 180 species such as apes, monkeys, chimpanzees, gibbons, macaques, mandrills, orangutans, baboons, lemurs, etc.

13. See the discussion later in this chapter concerning the scientific evidence for the first human couple living 200,000 years ago.

been more accurate than he imagined when he had all the animals calling each other "br'er," that is, "brother" or "kin." Those who have a hard time with calling monkeys their "cousins" have an even harder time with lions, tigers, rabbits, rats, and bears. Going back further still, humans and all animals (so the theory goes) have a common ancestor. Thus we would all be cousins to spiders, alligators, and slugs. Finally, if we go back far enough, all life on earth has evolved from simple one-cell creatures. These one-cell creatures developed from proteins and amino acids that formed on or in the primordial prebiotic earth or oceans. Presumably the proteins and amino acids were formed by the reaction of chemicals in the earth's waters or underground due to lightning discharges, intense heat, or other natural occurrences. As stated in the 1992 best-selling book *Shadows of Forgotten Ancestors, A Search for Who We Are* ("*Ancestors*") by Professor Carl Sagan and Ann Druyan:

> . . . *it seems very clear that there's only one hereditary line leading to all life now on Earth. Every organism is a relative, a distant cousin, of every other.*[14] *[Emphasis in the original]*.

Based on the theory of evolution, the development of humans can be understood as a natural phenomenon. There is no need for an almighty God creating the plants, animals, and humans in a cosmic week (as described in Genesis, the first book of the Bible). There is no need for humans to have supernatural souls that live on after death. Humans can be seen as simply one more biological organism whose members live and die, one more species that evolved over countless eons of time. According to evolutionary theory, human beings are a chance occurrence resulting from the random genetic mutations and genetic variations of previous animals and are no more special than any other species. Since, presumably, evolution continues even in the present, humans are not some ultimate divine development.

---

14. Sagan (1992), pg. 99.

In terms of the long evolutionary time-span between the original creatures and now, humans have existed only a relatively short period of time. Depending on long-term climatic changes and future random mutations, humans might or might not continue their prominent role in the world and could even become extinct. Mr. Darwin even opined:

> *Judging from the past, we may safely infer that not one living species will transmit its unaltered likeness to a distant futurity.*[15]

Some people fear that evolution could also have implications concerning the meaning of "right" and "wrong" and "good" and "bad." For example, the statement of Professor Provine excerpted earlier claims that there are no inherent moral or ethical laws and no guiding principles for humans. He claims that heredity, interacting with environmental influences, determines the ethical makeup of a human. Clarence Darrow, the defense attorney at the Scopes trial, used a similar argument in a murder trial as described in the previous footnote concerning Clarence Darrow. Others use similar reasoning to claim that in terms of evolution, an action is "good" if it helps ensure the survival of the species whose members perform the action. Thus, in their view, the fight to the death of two male animals of a certain species seeking to establish a territory or secure a female partner is "good" in evolutionary terms if it means that the genes of the stronger male will be passed on to the offspring. This will tend to make the offspring stronger and better able to defend themselves and stay alive until they in turn can produce their offspring.

For other species there may be more sophisticated interactions among the members of the species that allows territory and mates to be chosen without the death of weaker members but perhaps not without violence or the threat of violence. On the other hand, some claim that humans have the best chance of

---

15. Darwin (1859), p. 489.

surviving and prospering by renouncing violence and learning to treat each other with honor and respect. But, according to them, this would be "good" in evolutionary terms only if it ultimately results in the long-term survival of humans.

If the conclusions resulting from the theory of evolution concerning the origin of humans, the possible destiny of humans, and the concepts of good and evil are as described in the preceding paragraphs, the very core beliefs of most world religions are drawn into question. Some people fear that a completely natural explanation for the origin of humans discredits the belief in a supreme supernatural being (God). They fear that evolution implies that humans are just a chance development and not the culmination of God's creation. Rather than being "special children of God," they believe evolution relegates humans to being just another animal and possibly destined for extinction. This may be the reason some religious people find it hard to accept the theory of evolution. Others object to it because it seems to contradict the biblical account of creation in the book of Genesis. Still others believe that the evidence purported to support the theories of special creation and intelligent design brings into question some of the basic tenets of evolution.

On the other hand, there are many people, including scientists, who find no inherent conflict between the theory of evolution and their religious beliefs. Many of them believe that an almighty Supreme Being could use any mechanism, even evolution by natural selection, to have created humans, as well as all other biological organisms, over a long period of time. In addition, many of them recognize that evolution can be used to explain the development of the human body while a supernatural being is needed to explain the creation of each human soul that is miraculously integrated with each body. Many scientists, however, have difficulty accepting that a supernatural soul can interact with a material body. This issue is addressed in more detail in the chapter "The Soul-Brain Interface."

In *Ancestors*, the authors describe the rationale for the belief that the theory of evolution does away with the need for a divine being. For them, evolution presents the picture of humans as only one of millions of species that have evolved since the beginning of time. No direct hand of a divinity is needed. If a God is involved at all, it is only remotely, through setting the universe in motion billions of years ago. The only place God is needed, if even then, is at the very beginning of the universe, setting the big bang[16] in motion and letting the universe evolve over countless eons until galaxies, stars, and planets form. After more millions or billions of years, the earth cooled down through natural means and then lightning discharges, radiation, heat, or other natural occurrences caused amino acids and proteins—the building blocks of deoxyribonucleic acid ("DNA"), ribonucleic acid ("RNA"), and chromosomes—to form in the primordial waters, under the earth, or even—as proposed by some—in outer space. As Mr. Sagan and Ms. Druyan opine in *Ancestors*:

> *Evolution suggests that if God exists, God is fond of secondary causes and factotum processes: getting the Universe going, establishing the laws of Nature, and then retiring from the scene. A hands-on executive seems to be absent; power has been delegated. Evolution suggests that God will not intervene, whether beseeched or not, to save us from ourselves. Evolution suggests we're on our own—that if there is a God, that God must be very far away. This is enough to explain much of the emotional anguish and alienation that evolution has worked. We long to believe that there's someone at the helm.[17]*

The above characterization of God is similar to the view held by the Deists during the Enlightenment (also known as the Age of Reason) in the 1600s and 1700s. The Deists believed that the

---

16. The "big bang" is the cosmic explosion theorized by scientists to have occurred at the beginning of the universe. Scientific evidence indicates that the big bang occurred about 15 billion years ago.

17. Sagan (1992), pg. 67.

universe was like a giant clock and that God set the universe in motion at the beginning of time like a cosmic watchmaker. They believed, however, that God has had no further involvement with the universe or the lives of people. This, of course, is contrary to many religions (including Christianity) that teach that God does have direct involvement in peoples' lives.

Whether or not the above conclusions concerning evolution are valid, they nonetheless explain why there is so much controversy surrounding evolution. It appears, however, that the concept of "evolution" as described in the above excerpt is used to reach some unwarranted conclusions. The implicit assumption made by the authors of *Ancestors* is that evolution, as a mindless, impersonal force, is the only thing that can affect the universe. The underlying assumption is that God has not and does not affect the material universe or its inhabitants, either directly through natural means or indirectly through supernatural means.

However, there is nothing in the theory of evolution that *precludes* God from being involved in the universe either directly or indirectly. The theory of evolution does *not* "suggest that God" either has not or "will not intervene." The theory of evolution deals only with the natural world. It does not make any conjectures about the supernatural world. Even if we assume the theory of evolution accurately describes how natural life on earth developed, there are many ways a supernatural God could continue to influence human affairs, including at least the following methods:

Method 1. By taking on a human body, God could influence the world in a physical way. Christians, for example, believe that God took a human form and communicated to humans using a body similar to other human bodies. Admittedly this cannot be documented with scientific evidence, but nonetheless it is *not precluded* by the theory of evolution, as the above excerpt seems to imply.

Method 2. The theory of evolution does not preclude God from using supernatural means to influence human affairs. The previous excerpt from *Ancestors* appears to be based on the erroneous assumption that humans do not have supernatural souls. The assumption that humans do not have supernatural souls incorrectly limits how a supernatural God could influence humans. If humans do not have souls, then the way God can interact with the world is limited to direct physical actions. However, if humans do have supernatural souls, God can influence the world in supernatural ways. For example, by putting on the earth creatures that have self-awareness and free will, God has profoundly affected the course of events compared to a world in which all creatures operate solely based on instinct and evolutionary forces. Such creatures can use their free will to choose a course of action that might be different than that dictated by the forces of genetic mutation, genetic variation, and natural selection, which are the driving forces in the theory of evolution. These creatures would have the ability to choose good over evil and could create a world enriched by culture and learning.

Method 3. Another way a supernatural God could influence a world inhabited by creatures with supernatural souls would be for God to suggest ideas concerning right and wrong in the souls and minds of certain people so that these ideas can be made known to all humans. An example of this is the idea of a moral code defining right and wrong. Another example is the idea of love. The ideas embodied in "love" and "moral codes" have had a very important impact on the history of humankind and the world but the possibility that these ideas might have come from God via inspiration to prophets and holy people is not recognized in the aforementioned excerpt from *Ancestors*.

Method 4. The scenario painted in the previous excerpt from *Ancestors* does not recognize that a supernatural God would be able to influence humans in their everyday lives. Industry throughout the world spends exorbitant amounts of money on advertising based on the assumption that the minds of humans beings can be influenced to affect what they will choose to do and buy. The theory of evolution, even if it is accurate, does *not* preclude an almighty God from being able to influence people by planting suggestions in their minds and souls. These suggestions do not necessarily have to deal with ideas of cosmic importance such as right and wrong but could deal with specific choices a person might have relative to dealing with other people or pursuing life goals.

It might not be possible to provide scientific evidence that God has influenced and continues to influence the world in the ways described above. However, neither does the theory of evolution preclude such influence from occurring. It is not scientifically appropriate to simply assume that humans do not have souls and then assert there is a scientific basis for believing that God is unable to influence the world.

This narrow-mindedness regarding how God can operate in the world reminds me of the story of the man sitting on top of a house surrounded by floodwaters. He refuses to go with two boats and a helicopter that offer to take him as he exclaims that God will save him. After he drowns, he goes to heaven and asks why God did not save him from drowning. God replies that He sent two boats and a helicopter and that should have been more than enough.

The previous excerpt from *Ancestors* is further perplexing due to the apparent recognition by the authors that the theory of evolution does not preclude humans being made up of more than just their physical bodies. As stated in *Ancestors*:

> We "know" we are more than just a set of extremely complex computer programs. Introspection tells us that. That's the way it feels.[18]

---

18. Sagan (1992), pg. 166.

I believe they are correct. The chapter "Math" presents the mathematical evidence that our minds are able to solve more than just the computational problems that computers can solve. The chapter "Free Will Test" provides evidence that humans do have free will, something not possible in a computer program. But the authors of *Ancestors* do not address the implication that if we are more than just complex biological computers, we must be made of something nonmaterial. I believe this nonmaterial "something" is a supernatural soul which can explain our ability to solve non-computational problems and to exercise our free will.

If a person believes the premise that God does not influence the world, as described in the previous excerpt from *Ancestors*, and if such person believes, as some scientists do, that even the big bang could have happened without divine initiation, then there is no need for a divine being. As Carl Sagan, one of the authors of *Ancestors*, remarked in the forward to the 1988 edition of *A Brief History of Time* by Stephen Hawking, ". . . there is nothing for a Creator to do."[19] These types of conclusions, whether or not justified, have renewed the controversy as to whether or not science and religion are compatible. Such comments are reminiscent of Friedrich Nietzsche's claim in the 1880s that "God is Dead" and the "Is God Dead?" headline on the cover of the April 8, 1966 issue of *Time* magazine.

The theory of evolution might also appear to relegate people to being just another animal and not the culmination of God's handiwork. Man is able to claim a position as a very intelligent and resourceful animal, but an animal nonetheless. Mr. Darwin himself believed that man's intelligence is nothing more than the development of natural animal intelligence, quantitatively but not qualitatively different from other animals. As described in *Ancestors*:

> *"Most of the philosophers adjudged great in the history of Western thought held that humans are fundamentally*

---

19. Hawking (1988), 1988 ed., Foreword.

different from the other animals. Plato, Aristotle, Marcus Aurelius, Epictetus, Augustine, Aquinas, Descartes, Spinoza, Pascal, Locke, Leibniz, Rousseau, Kant, and Hegel were all proponents of the view that man differs radically in kind from [all] other things; except for Rousseau, they all held the essential human distinction to be our reason, intellect, thought, or understanding."[20] *Almost all of them believed that our distinction arises from something made neither of matter nor of energy that resides within the bodies of humans, but of no one else on Earth. No scientific evidence for such a "something" has ever been produced. Only a few of the great Western philosophers—David Hume, for instance—argued, as Darwin did, that the differences between our species and others were only of degree.*[21]

The "something" that is unique to humans as described in the above excerpt from *Ancestors*, is free will and the supernatural soul which is the subject of this book. This book provides a summary of the scientific evidence requested in the above excerpt. However, it is likely not possible to directly detect a supernatural soul using scientific instruments. Since the soul is a nonmaterial entity, scientific evidence is obtained in an indirect way. Scientific evidence is supplied by examining the effect that the soul has on material things (that is, human brains and bodies) rather than by directly detecting the soul. We have evidence that supernatural human souls exists because we can see that they affect material brains by allowing people to make free choices. As described by Professor Hawking in *A Brief History of Time,* and as excerpted in the chapter "Free Will or Not," identifying the existence of something by observing its effects (such as the effects of virtual particles) is a method that has long been accepted as valid by the scientific community.

---

20. Sagan (1992), pg. 364 as quoted from Mortimer J. Adler, *The Difference of Man and the Difference It Makes* (New York: Holt, Rinehart and Winston, 1967), p. 84.

21. Sagan (1992), pg. 364.

The evidence of human free will that leads to the conclusion that humans have supernatural souls returns humans to the center stage of creation. I believe that, due to such evidence, humans become something special in the universe; there is no evidence that other creatures anywhere on Earth or in the universe have free wills and supernatural souls. Humans regain their status as creatures at the center of importance in the universe, the culmination of God's handiwork, not because their bodies are created in God's image, but because their souls are created in God's image. I believe that supernatural souls give humans God-like powers: the power to make free choices, the power to love, the power to choose good over evil, the power to have ideas, the power to forgive, the power to create, and the power to feel joy. If humans are endowed with the special supernatural powers inherent in souls, I believe it is logical to conclude humans are not just some chance happening in the history of the universe but are very likely the main or even the only reason the universe was created.[22] If each human has free will and a supernatural soul, which are not natural phenomena and are not products of evolution, each human life becomes in a very real sense a miracle.

A Short Description of the Theory of Evolution

The theory of evolution is based on changes in the characteristics of organisms through genetic variation, genetic mutation, and natural selection. Natural selection is a theory of a process in which beneficial traits that enable an organism to survive and reproduce are passed on to succeeding generations by the combined effect of the following factors:

- Organisms naturally pass on their genetic code to their offspring in the reproductive process through DNA and/or RNA.

---

22. I believe the universe was created to give human bodies a place to live while their souls are finding the way to God. God created an awe-inspiring universe to demonstrate His power and majesty.

- Through the normal genetic process, the traits passed on to offspring can vary. Brothers and sisters, for example, are not normally identical and can have various traits inherited from their parents.
- Occasionally, there are genetic mutations that cause one or more traits to change. Some mutations can produce favorable traits, some unfavorable traits.
- A trait that is beneficial and makes it more likely for an organism to survive and reproduce in the specific environment in which the organism lives will have an enhanced probability of being passed on precisely because the trait gives the organism a greater probability of surviving and reproducing.
- A trait that is not beneficial and makes it less likely for an organism to survive and reproduce in the specific environment in which the organism lives will have a reduced probability of being passed on precisely because the trait gives the organism a lesser chance of surviving and reproducing.
- The result of the above-described process is an increase in the frequency of the favorable traits in the population of the species over time.

In the simplest terms, natural selection means that the traits of a species are "selected" by adaptation to the environment. In addition to physical traits, some of the traits determined by natural selection are behavior patterns. These behavior patterns operate based on complex decision processes.

THE ORIGIN OF SPECIES

What then is the origin of plant and animal species? How, according to the theory of natural selection, do species develop? Mr. Darwin noted that plants and animals varied. Although not known at the time of Mr. Darwin, it has since been shown that naturally occurring radiation is one source of genetic mutations. This is described in more detail below. Some theorize that the reproductive process also leads from time to time to random genetic mutations.

Most biologists believe that most of the genetic mutations that occur are not beneficial and do not help the organism survive and reproduce. Occasionally, however, a beneficial mutation occurs that does make it more likely for the organism to survive and reproduce better than the other members of its own species. Since this beneficial trait is embedded in the genetic code of the organism, it will have a chance to be passed on to the offspring of that organism. If the beneficial trait is passed on, the offspring will more likely survive and they will have a chance to pass on the trait to other members of the species. Eventually, over a very long period of time (thousands or millions of years), the trait will have a chance to spread throughout the entire population of the species, if there is no barrier between all members of the species.

On the other hand, there might be a natural barrier between two populations of one species that would prevent the two groups from intermingling and would prevent the various traits from intermingling. Thus, the mutations that occur and the traits that accumulate in one group would be different from the mutations that occur and the traits that accumulate in the other group. In this way, the two groups can evolve differently such that eventually they change enough genetically (again over very long periods of time) that they cannot interbreed (which is essentially the definition of separate species). The genetic differences between the two separate groups can be accelerated if the regions they occupy have different environments so that the mutations and traits that are most adaptable to one environment are different from the mutations and traits that are most adaptable to the other environment.

Evidence Supporting the Theory of Evolution

Some of the controversy surrounding the theory of evolution is due to the nature of the evidence that supports it. It is not possible to perform a laboratory experiment that would test millions of years of evolution in the same way that a scientist can test a law of chemistry or physics. Rather, scientists rely in part on the fossil record for evidence of evolution. The fossil records recovered to date, however, do not contain evidence for all of the transitional forms that would be needed to track the evolution of all of the characteristics of all of the species. Also, the theory of evolution does not have the predictive power of a law of chemistry or physics. The theory of evolution only predicts that change will occur. It does not predict the specific changes that will occur or the species that will evolve.

The evidence for evolution can be found in many biology text books and is described for nonbiologists by Christian biology professor Kenneth R. Miller in his 1999 book *Finding Darwin's God: A Scientist's Search for Common Ground Between God and Evolution*. Professor Miller finds no contradiction between a belief in God and a belief in evolution by natural selection. Professor Miller believes the scientific evidence overwhelmingly supports the acceptance of evolution both as an historical occurrence and as an established biological mechanism. Some of the evidence presented by Professor Miller includes:

- There are fossils of plants and animals in relatively defined geological layers (strata) in the earth. Each higher layer (which would have been deposited later than the lower layers) contains plants and animals with characteristics similar to the fossils of plants and animals in the lower layers but with modifications.[23]
- Each large geographical land mass contains plants and animals that are unique to that land mass. Each geographical land mass contains fossils of plants and animals that are similar to the fossils in other geological layers and to the living plants and animals in that geographical land mass but not to the fossils and

---

23. Miller (1999), pg. 61.

living plants and animals in other geographical land masses.[24]

- Certain islands near large geographical land masses contain plants and animals similar to the plants and animals on the nearby land mass and not to plants and animals on islands near other large geographical land masses. This occurs even though the environmental conditions on the islands are similar.[25]

- The lowest layers (strata) of geological deposits contain fossils of mosses and ferns but not of flowering plants. This indicates that flowers have evolved in more recent times.[26]

- Fossilized feces of dinosaurs contain evidence of extinct organisms common to the period of the dinosaurs and do not contain evidence of more recent organisms.[27]

- There are examples of currently living "ring species" which contain members that are geographically separated and are different enough to be separate species but the members on each extreme are linked by intermediates.[28]

- There are examples of field experiments under which natural selection has selected for certain characteristics of various species.

- There are functional and structural similarities among many species including humans and other primates.

- There are microbiological similarities between all organisms on earth. Living creatures (including humans) share nearly every aspect of biological existence with other living creatures. For example, a

24. Miller (1999), pg. 42.

25. Miller (1999), pg. 93.

26. Miller (1999), pg. 61.

27. Miller (1999), pg. 62.

28. Miller (1999), pg. 47.

human muscle, nerve, or bone cell is not particularly different from a cell taken from the same tissue of another mammal. Genetic instructions are encoded in all plants and animals (including humans) in the same language of DNA and RNA. Human genes transplanted into other animals, bacteria, and even plants function perfectly.[29]

- There are imperfections in the structure and function of human organs that could have been improved upon if they had been designed rather than evolved.[30]

- Laboratory experiments demonstrate the ability of DNA, RNA, and proteins to mutate to provide a specific function in a relatively short period of time.[31]

- Recent studies of living organisms document rates of genetic change that are at least 1,000 times faster than the supposedly fast-paced macroevolutionary events demanded by punctuated equilibrium (which is discussed below).[32]

- The recent development of strains of bacteria that are resistant to antibiotics is evidence of the power of natural selection to create new organisms with improved features.[33]

- There are examples of transitional creatures in the fossil record.[34]

- Dating the age of rocks by measuring the ratios of radioactive and nonradioactive elements indicates that the oldest rocks on earth approach an age of 4.5 billion years. This does not provide evidence that evo-

29. Miller (1999), pg. 59.

30. Miller (1999), pg. 101.

31. Miller (1999), pg. 106.

32. Miller (1999), pg. 111.

33. Miller (1999), pg. 104.

34. Miller (1999), pg. 119, 124–125, 138–139 and 264–265.

lution by natural selection has occurred, but it does provide evidence that there has been enough time for evolution to have occurred.[35]

Mr. Darwin recognized that examples of transitional forms between species would help support the theory of evolution. These transitional forms would be organisms that have some of the traits of one species and some of the traits of another species. He cautioned, however, that these organisms would be transitions between the existing species and some common ancestor and not transitions directly between two existing species. Mr. Darwin explained that existing transitional forms are rare because they would have become extinct, having been replaced by the current, more advanced species. However, he theorized that the gradual development of all the existing species through small mutations over very long periods of time would have required an enormous number of transitional forms and that the fossil record should be full of such transitional forms. He also recognized that at the time he offered his theory in 1859, the fossil record did not provide a lot of evidence of transitional forms. He further theorized that this was simply due to the fact that the fossil record had not yet been fully developed. As he stated in *The Origin of Species*:

> *But as by this theory [of evolution by natural selection] innumerable transitional forms must have existed, why do we not find them embedded in countless numbers in the crust of the earth? . . . I believe the answer mainly lies in the record being incomparably less perfect than is generally supposed ....*[36]

> *Why then is not every geological formation and every stratum full of such intermediate links? Geology assuredly does not reveal any such finely graduated organic chain; and this, perhaps, is the most obvious and serious objection which can be urged against my theory [of evolution by natural selection].*

---

35. Miller (1999), pg. 69.
36. Darwin (1859), pg. 172.

The explanation lies, as I believe, in the extreme imperfection of the geological record.[37]

In an attempt to provide evidence that evolution by natural selection has occurred over time, scientists since the time of Mr. Darwin have searched for fossils that would indicate there were transitional forms. After searching for over 130 years and cataloging 250 million fossils, some scientists believe there is little, if any, evidence of transitional forms to support the theory of evolution by natural selection. As described in the 1994 book *The Creation Hypothesis*:

> *Yet, 130 years later, it has become clear that the fossil record does not confirm Darwin's hope that future research would fill in the extensive gaps in the fossil record. Paleontologist Stephen Jay Gould of Harvard points out, "The fossil record with its abrupt transitions offers no support for gradual change. . . . All paleontologists know that the fossil record contains precious little in the way of intermediate forms; transitions between major groups are characteristically abrupt."*[38]

> *With an estimated 250 million, or ¹/₄ billion cataloged fossils of some 250,000 fossil species, the problem does certainly not appear to be an imperfect record. Many scientists have conceded that the fossil record is sufficiently complete to provide an accurate portrait of the geologic record. University of Chicago Professor of Geology David Raup says, "Well, we are now about 120 years after Darwin, and the knowledge of the fossil records has been greatly expanded. We now have a quarter of a million fossil species, but the situation hasn't changed much. The record of evolution is still surprisingly jerky and, ironically, we have even fewer examples of evolutionary transition than we had in Darwin's time."*[39]

---

37. Darwin (1859), pg. 280.

38. Moreland (1994), pg. 278 as quoted from Stephen Jay Gould, "The Return of Hopeful Monsters," *Natural History*, June/July 1977, pp. 22, 24.

39. Moreland (1994) pp. 278-279, quoting from David Raup, "Conflicts between Darwin and Paleontology," *Field Museum of Natural History Bulletin*, January 1979, pp. 22, 25.

The Creation Hypothesis provides a number of similar observations by other prominent scientists. It should be noted, however, that as described above, Professor Miller has given examples of transitional creatures in the fossil record. It should also be noted that in spite of the above excerpt from Professor Steven Jay Gould, he was a fervent supporter of the theory of evolution. As described over the next few pages, he claimed to modify the theory to recognize a process he described as punctuated equilibrium.

Conclusions by scientists as previously described have given new impetus to the demands to require that students be provided with information on the claimed shortcomings of evolution and the evidence for the competing theories of special creation and intelligent design. A good explanation of the evidence comparing evolution and intelligent design can be found in the 1989 book *Of Pandas and People: The Central Question of Biological Origins* by Percival Davis and Dean Kenyon. Also, for a sampling of the views of a number of prominent authors on these matters, consult the 2001 book *Intelligent Design, Creationism, and Its Critics: Philosophical, Theological, and Scientific Perspectives.*[40]

THE CAMBRIAN EXPLOSION

In his 1997 book, *The Science of God*, Professor Gerald Schroeder described how in 1909 Charles Walcott, the director of the Smithsonian Institute in Washington, D.C., accumulated over 60,000 fossils from the Burgess Shale in British Columbia, Canada. These fossils were from the Cambrian period of 500 to 600 million years ago, and he collected representatives of every animal phylum. The fossils provide evidence that there had not been a gradual evolution of species but that life forms with various types of organs and traits appeared abruptly (in geological time frames). Mr. Walcott stored the fossils in drawers in the Smith-

---

40. Pennock (2001).

sonian where they remained until they were rediscovered in the mid-1980s. As Professor Schroeder described, the Burgess fossils appeared to change everything:

> *Their effect has been so dramatic that the most widely read science journal in the world, **Scientific American**, in its November 1992 issue was moved to question: "Has the mechanism of evolution altered?" The same reaction to these fossils appeared in the October 1993 issue of **National Geographic** magazine. The science section of the **New York Times** referred to the fossils as demonstrating "revolution more than evolution."[41] And **Time** magazine featured them in a comprehensive and scientifically accurate cover story titled "Evolution's Big Bang."[42]*

The Cambrian explosion is consistent with punctuated equilibrium, a term popularized by Jay Gould and Niles Eldredge. Punctuated equilibrium means that change occurs quickly (i.e., is punctuated) with species arising suddenly, followed by long periods without change (i.e., equilibrium or stasis). This is not consistent with evolution based only on continuous and gradual change over long periods of time. Professor Miller, however, points out that the theory of punctuated equilibrium is consistent with the Darwinian theory of evolution by natural selection. Although the periods of rapid (punctuated) change are short in geological time frames, they are long in terms of the time needed for evolutionary change and natural selection to have occurred. He notes that recent studies of living organisms document rates of genetic change that are "easily 1,000 times faster than the supposedly fast-paced macroevolutionary events demanded by punctuated equilibrium."[43] He also notes that Charles Darwin himself

---

41. Schroeder (1997), pg. 39, quoting from J. Wilford, "Spectacular Fossils Record Early Riot of Creation," *New York Times*, 23 April 1991, C-1.

42. Schroeder (1997), pg. 39, ref. M. Nash, "When Life Exploded," *Time*, 4 December 1995.

43. Miller (1999), pg. 111.

believed in punctuated equilibrium although he never used that terminology. Mr. Darwin believed that the process of evolution would be "irregular," that forms would remain unaltered for long periods and then again undergo modification.[44] Note also that the punctuated modifications that occurred quickly (in geological time frames) might have contributed to organisms leaving a lot fewer fossils of transitional forms than might have been deposited in the geological record had the changes occurred gradually over longer periods of time.

It should be noted that scientists are virtually unanimous in their acceptance of evolution by natural selection as a robust theory that explains a lot of evidence accumulated over the years. Many books have been written to respond to the evidence that supports the theories of special creation (a.k.a "creation science") and intelligent design. In his 1999 book *Finding Darwin's God*, Professor Miller presents some of the evidence and arguments that refute the theories of special creation and intelligent design. In an article in the July 2002 issue of *Scientific American* entitled "15 Answers to Creationist Nonsense," editor John Rennie responds to some of the issues raised by the theories of special creation and intelligent design. He also provides information on other sources that describe the evidence supporting the theory of evolution. I suggest all interested readers should review Mr. Rennie's article concerning the available evidence. In spite of the condescending title of his article, I believe that fully (and respectfully) considering the scientific evidence will ultimately lead to the truth about evolution and other theories.

It is irrelevant to me whether or not evolution correctly explains how plant and animal species (especially humans) developed. Based on the biblical description, human bodies are made out of the dust of the earth.[45] Based on the theory of evolution, human bodies are descended from one-celled animals and are made out

---

44. Miller (1999), pg. 113–115.

45. Genesis 2:7.

of materials from the earth. Neither description is very gratifying in and of itself. Human worth is elevated not by how our bodies are made but by the supernatural soul that each human has. I am hopeful that both scientists and religious people can agree that human free will cannot be explained as a natural phenomenon that is subject to the laws of physics and chemistry (which is consistent with the beliefs of William Provine and Charles Darwin). I am hopeful that they can then agree that this leads to the conclusion that the source of human free will is a supernatural phenomenon. I hope this will reduce the conflict associated with the question of whether human bodies are the product of gradual evolution, punctuated evolution, or direct design by a Creator.

THE ORIGIN OF LIFE

It is important to remember that the theory of evolution by natural selection requires life to already be existing. The theory of evolution purports to explain the origin of species. It does not purport to explain the origin of life itself. Mr. Darwin's theory of evolution required that there already be in existence one or more life forms into which life had been breathed. From these original life forms "endless forms most beautiful and most wonderful have been, and are being evolved." He believed that all the facets that make up the theory of evolution provide a grandeur to life, and he viewed life as being ennobled by evolution. As he described in the final pages of his book *On the Origin of Species:*

> *Authors of the highest eminence seem to be fully satisfied with the view that each species has been independently created. To my mind it accords better with what we know of the laws impressed on matter by the Creator, that the production and extinction of the past and present inhabitants of the world should have been due to secondary causes, like those determining the birth and death of the individual. When I view all beings not as special creations, but as the lineal descendants of some few beings which lived long before the first bed of the Silurian [Cambrian] system was deposited, they seem to me to*

become ennobled. Judging from the past, we may safely infer that not one living species will transmit its unaltered likeness to a distant futurity. . . . There is a grandeur in this view of life, with its several powers, having been originally breathed [by the Creator] into a few forms or into one; and that, whilst this planet has gone cycling on according to the fixed law of gravity, from so simple a beginning endless forms most beautiful and most wonderful have been, and are being evolved.[46] *[Brackets indicate words added by Mr. Darwin in the second and subsequent editions.]*

As indicated above, Mr. Darwin did not know who or what originally breathed life "into a few forms or into one." His theory only purported to describe how the numerous species in the world developed from the original few life forms. In the second addition of his book *On the Origin of Species*, he was even willing to add the words "by the Creator" to indicate the source of the breath of life. This was done to help thwart opposition from religious critics. As described by Janet Browne in her 2002 book *Charles Darwin, The Power of Place*, Mr. Darwin hoped that by providing "the Creator" with a hand in the development of life, he could diffuse the religious objections to his theory and win the endorsement of sympathetic clergymen such as Reverend Charles Kingsley, an Anglican clergyman and popular author.[47] He also did not want to alienate religiously devout scientific friends, like the botanist Asa Gray. This he hoped would allow people to focus on the science of his theory about how species developed and not be concerned about any apparent religious or spiritual implications. In the chapter on "Instinct," Mr. Darwin confirmed that his theory did not explain the origin of life:

I must premise, that I have nothing to do with the origin of the primary mental powers, any more than I have with that of life itself. We are concerned only with the diversities of

---

46. Darwin (1859), pp. 488-490.

47. Browne (2002), Volume 2, pp. 95-96.

instinct and of the other mental qualities of animals within the same class.[48]

Mr. Darwin did not have the scientific evidence he would need to develop a theory about the origin of life and was not aware of the complexity associated with the genetic codes, DNA and RNA. About 90 years later, in the 1950s, biologists James D. Watson of the United States and Francis H. C. Crick of Great Britain discovered the chemical structure of DNA and RNA which have been called the molecular blueprint of life. DNA and/or RNA make up genes which contain all the hereditary information that regulates the characteristics of an organism. Genes are located in chromosomes which can be found in the nucleus of each cell in the organism. DNA and RNA are made out of amino acids that are arranged in very complex sequences to regulate all the life functions in all living organisms.

In 1952, Stanley Miller performed an experiment in which amino acids were formed in laboratory tubes by discharging an electrical spark (to simulate lightning) in what was considered to be early earth atmospheric conditions of ammonia, methane, hydrogen, and water vapor. As described in the 1992 book *The Creation Hypothesis*:

> *A spate of follow-up experiments simulating early-earth conditions have subsequently produced nineteen of the twenty biological amino acids (only lysine remains), all five nucleic acid bases found in RNA and DNA, and various fatty acids found in cell membranes.*[49]

It appeared that scientists would soon be able to provide evidence that life itself could develop on its own from purely natural occurrences. But the task has proved to be much harder than originally expected. The primary reason it is so difficult is that DNA and RNA are not random arrangements of amino acids but

---

48. Darwin (1859), pg. 207.
49. Moreland (1994), p. 182.

are very complex arrangements that contain incredible amounts of information, much like computer code. Amino acids are only chemical compounds. They are not living creatures. By themselves, amino acids are *not* complex arrangements of information that can regulate how an organism grows and reproduces. Simply throwing amino acids together into the same mixture and swirling them together, whether for one day or 5 billion years (the age of the Earth as generally accepted by scientists), will not easily result in the complex structure which makes up DNA and RNA. Some scientists have estimated that even if all carbon on earth existed in the form of amino acids, there is essentially zero probability of getting amino acids to form together into one functional protein by means of random reactions over a few billion years.[50] Scientist Sir Frederick Hoyle best described the improbability of this occurring when he opined, in his 1983 book *The Intelligent Universe,* that the origin of life by random reactions of amino acids (as currently envisioned by some scientists) was about as likely as the assembly of a Boeing 747 jet airplane by a tornado whirling through a universe of junkyards, each of which contained all of the pieces of the 747 jet airplane.[51] In his 1991 book *Darwin on Trial,* Phillip Johnson opined:

> *Chance assembly is just a naturalistic way of saying "miracle."*[52]

The difficulties that have to be overcome to be able to explain the origin of life based on random chemical reactions are described in *Darwin on Trial* and in more comprehensive scientific detail in *The Creation Hypothesis.*[53] Some of the items covered include:

---

50. Moreland (1994), p. 190.

51. Hoyle (1983), pg. 19. In the same book, Professor Hoyle described a theory of panspermia, that the earth was seeded with micro-organisms from space. In my opinion, there would likewise be a similar low probability for micro-organisms developing through random reactions in interstellar space. His only explanation for the source of these micro-organisms appeared to be that they were created by an Intelligent Universe.

52. Johnson (1991), pg. 104.

53. See also the 1999 book *The Fifth Miracle: The Search for the Origin and Meaning of Life* by Paul Davies.

- The currently accepted composition of the gases of the early-earth atmosphere is not conducive to the production of amino acids, the building blocks of life.
- The synthesis of key building blocks of DNA and RNA has never been accomplished in the laboratory except under highly implausible conditions without any resemblance to those of early earth.
- There is enormous difficulty in achieving the very specific forms and chemical bonds required for functioning proteins.
- The pathways which are needed to synthesize DNA or RNA from proteins are extremely complicated.
- Scientists do not yet have a solution to the chicken and egg problem of which came first, DNA or RNA.
- Scientists do not yet have a plausible description of how natural selection would work on inanimate matter.

Note, however, that as to the first item listed above, some scientists have theorized that life may have originated far underground or deep in the oceans near volcanic vents. This theory is supported by the discovery of microbes that can live in extremely hot conditions and have a metabolism based on nonorganic mineral nutrients and driven by thermal energy rather than by sunlight like surface plants that rely on photosynthesis.[54]

Note that a discovery of a natural explanation for the origin of life would not demonstrate that free will can be explained as a biological mechanism. They are two entirely different matters.

INTELLIGENT DESIGN

The above discussion relative to the difficulties associated with the development of DNA and/or RNA based on random combinations of amino acids leads to a consideration of the theory of intelligent design. Intelligent design theory is based on

---

54. See, for example, Davies (1999).

the hypothesis that there are certain developments that are so complex that they cannot be explained by natural causes. This implies that the only reasonable explanation is that they are the work of an intelligent designer. In the 1998 book *Mere Creation: Science, Faith, & Intelligent Design*, William Dembski[55] describes a test for identifying whether or not an object or mechanism was developed by an intelligent designer. He described three possible explanations for a given development: random occurrence, physical forces (as described by physical laws), and intelligent design. An example of random occurrence is the arrangement of grains of sand on a beach. An example of physical forces is the arrangement of iron filings in a magnetic field.

An example of intelligent design is the arrangement of large stones called Stonehenge in England. The arrangement of Stonehenge does not appear to be due to the random movement of stones. There is likewise no known physical force that could explain the arrangement. The only reasonable explanation for the Stonehenge monument is that the stones were arranged by an intelligent designer, even though we do not know who arranged them, when they were arranged, or why they were arranged. Scientists use the concept of intelligent design in archaeology, the scientific study of the life and culture of ancient peoples through the excavation of ancient cities and the inspection of relics and artifacts. The relics and artifacts have characteristics that allow scientists to recognize that they were developed by an intelligent designer.

---

55. William Dembski is associate research professor in the conceptual foundations of science at Baylor University and senior fellow with the Discovery Institute's Center for the Renewal of Science and Culture in Seattle. A graduate of the University of Illinois at Chicago, where he earned a B.A. in psychology, an M.S. in statistics, and a Ph.D. in philosophy, he also received a doctorate in mathematics from the University of Chicago in 1988 and a master of divinity degree from Princeton Theological Seminary in 1996. He has held National Science Foundation graduate and postdoctoral fellowships. Dr. Dembski has published articles in mathematics, philosophy, and theology journals and is the author of seven books.

The United States government has authorized millions of dollars to search for evidence of radio transmissions from outer space in hopes of discovering an intelligent message. Clearly there is a common sense as well as a scientific basis for identifying when something has been developed through intelligent design.

As described previously, scientists have not yet been able to provide evidence that DNA and/or RNA have developed as a result of random occurrences or physical forces. Some scientists believe the complexities of DNA, RNA, and certain cellular structures and functions exhibit the characteristics of something developed by intelligent design. In essence, the theory of intelligent design asks the question: What kind of evidence leads to the conclusion that something has occurred randomly or through natural selection and what kind of evidence leads to the conclusion that something has occurred due to the efforts of an intelligent designer?

In his 1996 book *Darwin's Black Box*, Michael Behe[56] describes complex structures and functions in the biomolecular systems of certain life forms. These biomolecular systems have interdependent parts that he claims cannot be explained as having evolved independently. Examples of such structures include cilia, bacteria flagellum, blood clotting, and the immune system. In *Finding Darwin's God*, Professor Miller rebuts Professor Behe's claims and provides evidence that the interdependent parts first evolved to serve other functions but were opportunistically used by the organisms to provide the complex functions identified by Professor Behe. Professor Behe replies that the examples given by Professor Miller are only generic and do not provide evidence as to how the specific modifications occurred to produce the complex cellular functions he identified.[57] This is an area that will likely remain controversial for the foreseeable future.

---

56. Michael Behe is associate professor of biochemistry at Lehigh University.

57. Current, ongoing discussions concerning these issues can be found at *www. miller.org*, *www.crsc.org* and *www.intelligentdesignnetwork.org*.

In his 2002 book *No Free Lunch: Why Specified Complexity Cannot Be Purchased without Intelligence,* William Dembski, relies on "advances in probability theory, computer science, the concept of information, molecular biology, and the philosophy of science" to conclude (1) that there are parts of the biological world that exhibit complex specified information and (2) that natural causes (including natural selection) cannot explain such complex specified information. Professor Dembski described the basis for his book:

> *The great myth of contemporary evolutionary biology is that the information needed to explain complex biological structures can be purchased without intelligence. My aim throughout this book is to dispel that myth.*[58]

Professor Dembski claims that the creation of specified information in biological systems requires an intelligent agent and that the ability to make choices is critical to exercising that intelligence to create information.

> *The principal characteristic of intelligent agency is choice. . . . For an intelligent agent to act is therefore to chose from a range of competing possibilities.*[59]

Professor Dembski explains that natural systems that are subject either to deterministic or probabilistic laws do not have the freedom to make the choices needed to create the complex specified information present in biological systems. This is very similar to the basic premise of this book that free will cannot be explained by natural systems that are subject either to deterministic or probabilistic laws. Professor Dembski also claims that the universe has not existed long enough for there to be a reasonable chance that the complex specified information in biological systems occurred due to chance. An example of this is provided later in this chapter.

---

58. Dembski (2002), pg. 148.
59. Dembski (2002), pg. 28.

Professor Dembski would like scientific criteria to be applied consistently when used to identify which phenomena are produced by random occurrences and which phenomena are produced by intelligent agents. As described later in this chapter, one scientist has calculated that there is an incredibly low probability that the proteins needed for life occurred due to chance. If radio signals from outer space were to be detected in complex patterns that exhibit the same low probability of being a random occurrence,[60] would scientists proclaim that they represent evidence of an extraterrestrial intelligence? Would scientists apply the same criteria to the development of the proteins needed for life and likewise describe them as evidence of an intelligent designer?

One additional related question concerns the capability of human beings to be intelligent designers. Intelligent design means that the designer uses foresight to develop the parts of a mechanism and arrange them in a certain pattern to perform a specific function. Is it logically possible for a process such as evolution through natural selection (which does not operate through foresight or planning) to produce a creature that has the capability of intelligent design (which does operate through foresight and planning)? Professor Demski concludes that a being made up only of natural phenomena that are subject to deterministic and probabilistic laws does not have the freedom to make the choices that intelligent design requires. Based on this, it is logical to conclude that no natural process could produce a creature capable of intelligent design. As Professor Dembski explains:

> *I will argue that intelligent agency, even when conditioned by a physical system that embodies it, cannot be reduced to natural causes without remainder. Moreover, I will argue that specified complexity is precisely the remainder that remains unaccounted for. Indeed, I will argue that the defining feature of intelligent causes is their ability to create novel information and, in particular, specified complexity.*[61]

---

60. Such as occurred in the movie *Contact*.

61. Dembski (2002), pg. xiv.

There are a couple of additional issues addressed in Professor Dembski's book on which I would like to comment. In the section in which he explains that choice is a critical characteristic of intelligence, he describes the choices made by a rat in a maze as being intelligent choices. I do not think the kind of creative intelligence he is investigating in *No Free Lunch* is the same kind of intelligence that a rat in a maze exhibits. The rat did not create the maze. The rat does not define the criterion that describes when the maze has been successfully navigated. It appears the rat is only driven by natural instinct to find the food at the end of the maze.

The second issue relates to the conclusion that the presence of complex specified information in biological systems requires there to be an intelligent designer. Since there were no humans around before there were biological systems, the presence of complex specified information implies to some people that there was an unembodied intelligent being that created the complex biological systems. He then opines that such an intelligent being could take advantage of the probabilistic nature of quantum mechanics and would not need to expend energy to implement the complex biological systems:

> ... *unembodied designers who co-opt random processes and induce them to exhibit specified complexity are not required to expend any energy.*[62]

Professor Dembski does not explain how the biological systems can be "induced" to exhibit specified complexity without the need to expend any energy. His position appears to be similar to that of Sir John Eccles as described in the chapter "The Soul-Brain Interface." I, however, do not agree that information can be induced into a system without expending energy.

---

62. Dembski (2002), pg. 341.

WHAT IS LIFE?

The above discussion focuses our attention on the central question of biology: What is life? As alluded to above, living organisms are conceptually different from other combinations of elements and molecules found on the earth. Ordinary matter is subject to the laws of physics, but the actions of living organisms have another layer of control: DNA and/or RNA. Although DNA and RNA are also subject to the laws of physics, there is information encoded in DNA and/or RNA that regulates organisms concerning how to function and how to reproduce. (And as explained throughout this book, I believe human life has yet another layer of control that is *not* subject to the laws of physics: free will via a supernatural soul.)

In his fascinating 1999 book *The Fifth Miracle: The Search for the Origin and Meaning of Life*, physicist Paul Davies describes how it is the information encoded in DNA and RNA that makes life different from other matter in the universe. As he and others have noted, DNA and RNA are the blueprints of life. All living organisms, including all plants and animals, have functioning DNA and/or RNA. This leads to the following definition of life (on earth): A living organism is anything that has functioning DNA and/or RNA.

As investigated in great detail in Professor Dembski's book *No Free Lunch*, the presence of such complex specified information leads to the following question: Where did the information in the DNA and RNA come from? How can atoms and molecules randomly following the laws of physics come to be in such a complex structure which contains so much information? As Professor Davies notes, even if Darwinian natural selection is a valid explanation for the development of species, it does not explain where DNA and/or RNA came from in the first place.

> *Darwinism can offer absolutely no help in explaining that all-important first step: the origin of life.*[63]

---

63. Davies (1999), pg. 44.

> *What remains to be explained—what stands out as the*
> *central unsolved puzzle in the scientific account of life—is*
> *how the first [self-replicating] microbe came to exist.*[64]

As described by Professor Davies in *The Fifth Miracle*, scientists are very far from having a plausible explanation as to how the complex, information-rich structures and systems needed for life could have developed from the molecules and atoms that existed in the primordial Earth. And even if a plausible explanation is found, it has to receive further verification by experimental evidence before it can be accepted as a valid scientific explanation. Professor Davies describes the exceedingly low probability that the formation of proteins needed for life could have occurred by chance:

> *In the previous section, I presented the fantastic odds*
> *against shuffling amino acids at random into the right*
> *sequence to form a protein molecule by accident. That was a*
> *single protein. Life as we know it requires hundreds of thou-*
> *sands of specialist proteins, not to mention the nucleic acids.*
> *The odds against producing just the proteins by pure chance*
> *are something like $10^{40,000}$ to 1. This is one followed by forty*
> *thousand zeros,*[65] *which would take up an entire chapter of this*
> *book if I wanted to write it out in full. . . .*
>
> *There are indeed a lot of stars—at least 10 billion bil-*
> *lion in the observable universe. But this number, gigantic as*
> *it may appear to us, is nevertheless trivially small compared*
> *with the gigantic odds against the random assembly of even*
> *a single protein molecule. Though the universe is big, if life*

---

64. Davies (1999), pg. 29.

65. It is very difficult to explain how large this number is. It is one billion billion billion billion, etc. (with the word "billion" repeated 4,444 times). This means one billion multiplied by itself 4,444 times. By contrast, the number of stars in the observable universe is 10 billion billion.

formed solely by random agitation in a molecular junkyard,[66] *there is scant chance it has happened twice.*[67]

In spite of these odds and without having any definitive evidence to support his explanation, Professor Davies confidently proclaims:

> *Though I have no doubt that the origin of life was not in fact a miracle, I do believe that we live in a bio-friendly universe of a stunningly ingenious character.*[68]

> *Nobody wrote the message; nobody invented the [DNA and RNA] code. They came into existence spontaneously. Their designer was Mother Nature herself, working only within the scope of her immutable laws and capitalizing on the vagaries of chance.*[69]

Such statements as the ones above, without any evidence to support them, are based only on "belief" and "faith" which are supposed to be the realm of religion, not science. In fact, the evidence concerning the odds against the proteins needed for life developing by the "vagaries of chance" would seem to point to the opposite conclusion. With such tremendous odds against the proteins needed for life occurring by random reactions of amino acids, the most reasonable explanation at this point in the history of science would be that the components needed for life were developed by an intelligent designer or some unknown natural process.

In his book *No Free Lunch*, Professor Dembski explains that there is a limit as to how many nuclear interactions could have

---

66. This is a reference to the comment of Fred Hoyle comparing the random development of DNA/RNA to the assembly of a Boeing 747 jet airplane by tornadoes, which was described previously in this chapter.

67. Davies (1999), pg. 95.

68. Davies (1999), pg. 20.

69. Davies (1999), pg. 41.

occurred during the period the universe has existed. It is estimated there are 10^{80} subatomic particles (electrons, protons,
and neutrons) in the universe (this is a 1 followed by 80 zeroes).
According to quantum theory, the maximum number of interactions that an atomic particle can undergo is 10^{45} per second.[70]
The universe has existed for 10^{25} seconds. By multiplying these
three numbers together (by adding the exponents), we find that
there could have been at most 10^{150} subatomic interactions during
the period the universe has existed. This means that any activity
that requires more than 10^{150} interactions could not have occurred
during the time the universe has existed. This is an incredibly
large number of interactions (1 followed by 150 zeroes). Nevertheless, it is infinitesimally small compared to the estimate of the
number of interactions required for the formation of the proteins
needed for life to occur, on average, once by chance as described
above by Professor Davies. One chance in $10^{40,000}$ means that for
every $10^{40,000}$ interactions of particles, the proteins would form
once, on average, due to chance. Based on the maximum number
of interactions that could have occurred during the existence of
the universe, there would have to be a period of time that is at
least $10^{39,850}$ times as long as the universe (a 1 followed by 39,850
zeroes) to provide enough time for $10^{40,000}$ interactions to have
occurred. If the estimate of 1 in $10^{40,000}$ is within several thousand orders of magnitude of the actual probability that proteins
developed by chance, it is clear that the evidence does not support the hypothesis that the formation of the proteins occurred
by chance.

Professor Dembski also addresses the claim that the proteins
could have developed by chance based on the theories that there
are many other universes that exist that could have allowed the
required large number of interactions to have occurred. He notes,

---

70. The shortest period of time is 10^{-45} second, which is known as Planck's time.

however, that there is no evidence to support the theories of many other universes and that it is unscientific to rely upon such theories to explain anything until there is supporting evidence.

In my opinion, the limitations relative to the ability of science to explain the origin of life as the product of random interactions of particles as described above should be taught whenever theories about the origin of life and the evolution of species are taught. Professor Davies also describes the theory that life might have originated on Mars and migrated to Earth in meteorites blasted from the Martian landscape by large asteroids that crashed into the Martian surface. This theory does not solve the problem of the complexity of DNA and RNA. Regardless whether DNA and/or RNA originated on Earth or Mars (or somewhere else in the universe),[71] to provide a scientifically supported hypothesis that life developed spontaneously without the intervention of an intelligent designer, scientists must provide experimental evidence for life developing under conditions existing somewhere on Earth or Mars (or even some hypothetical conditions somewhere else in the universe).

The discovery of DNA and RNA as the blueprint and indicators of life provides insight into the beginning of human life. The evidence indicates that new human life begins at the moment of conception when the DNA and RNA that regulates that particular human life is formed. The life formed with the DNA and RNA code at the moment of conception is human: such life has a human mother and a human father. The DNA and RNA in that human life is different from the DNA and RNA of the father and is different from the DNA and RNA of the mother. Although the human life inside the mother takes nourishment from the mother's body, it has different DNA and RNA and is not a part of the mother's body.

---

71. The concept that the universe might have been "seeded with life" in a process of panspermia has been hypothesized by Fred Hoyle and Francis Crick, one of the co-discoverers of DNA.

THE SIX DAYS OF CREATION

One of the original sources of contention between those who believed in creation and those who believed in evolution was the length of time that was required for all of the plants and animals to come into existence. The first chapter of the book of Genesis in the bible appears to describe a six day period during which the Creator created the universe, the sun, the moon, the Earth, and all the plants and animals on the Earth. There is scientific evidence, on the other hand, that the universe is about 15 billion years old (give or take a few billion years) and that the Earth is 4 to 5 billion years old. With incredible insight, Professor Schroeder, who has a Ph.D. in physics, describes a way to reconcile these two seemingly irreconcilable positions using Einstein's theory of relativity.

According to Einstein's theory of general relativity, time slows down near areas of high concentrations of mass and energy.[72] This feature of time is confirmed by Professor Stephen Hawking in his popular book *A Brief History of Time*:

> *Another prediction of general relativity is that time should appear to run slower near a massive body like the Earth. This is because there is a relation between the energy of light and its frequency (that is, the number of waves of light per second): the greater the energy, the higher the frequency. As light travels upward in the Earth's gravitational field, it loses energy, and so its frequency goes down. (This means that the length of time between one wave crest and the next goes up.) To someone high up, it would appear that everything down below was taking longer to happen. This prediction was tested in 1962, using a pair of very accurate clocks mounted at the top and bottom of a water tower. The clock at the bottom, which was nearer the Earth, was found to run slower, in exact agreement with general relativity. The difference in the speed of clocks at*

---

72. Remember that according to the equation $E=MC^2$ mass and energy are equivalent. They are different forms of the same thing.

different heights above the Earth is now of considerable practical importance, with the advent of very accurate navigation systems based on signals from satellites.[73]

In his 1977 book, *The First Three Minutes*, Professor (and Nobel prize winner) Steven Weinberg explains that shortly after the big bang beginning of the universe, the entire mass and energy of the universe was located at one very confined spot with a density of 3.8 billion kilograms per liter[74] (3.8 million metric tons per liter or 15.8 million short tons per gallon). This is 3.8 billion times as dense as water. This incredible amount of mass and energy would cause time to slow down immensely compared to equivalent Earth time.

Professor Schroeder defined a "cosmic" time that is measured by the cosmic background radiation (CBR). CBR is light that has been present since the beginning of the universe. CBR is detected from all directions in the universe and is often referred to as the "echo of the big bang." Professor Schroeder hypothesizes that the days referred to in the book of Genesis were days of cosmic time because the actions described were cosmic in nature.

His theory is that at the beginning of the universe, the cosmic time was going billions of times slower than equivalent Earth time due to the density of the universe and relativistic time dilation as described above. As the universe continued to expand after the big bang, the density of the universe dropped and the "cosmic" time began to speed up relative to Earth time. Due to the nonlinear decrease in density as the universe expanded, the speed-up of time was also nonlinear. Because of this nonlinear change, each succeeding cosmic day was equivalent (in Earth time) to one-half of the equivalent earth time of the previous cosmic day. The table below compares cosmic time to Earth time, starting with day one after the big bang and ending with day six. Note that cosmic days are 24-hour days.

---

73. Hawking (1988) 1996 ed., pg. 33-34.
74. Weinberg (1977), p. 103.

Table 4.1

Cosmic Time (in 24-hour days)	Equivalent Earth Time (in years)	Developments during the Period
Day 1	8 billion years	The creation of the universe, light separates from darkness
Day 2	4 billion years	The heavens form including the stars, the sun and the Earth
Day 3	2 billion years	Oceans and dry land appear, simple plant forms appear
Day 4	1 billion years	Earth's atmosphere becomes transparent, sun, moon and stars appear
Day 5	$1/2$ billion years	First animal life, reptiles and winged creatures
Day 6	$1/4$ billion years ending about 6,000 years ago	Land animals, mammals, humans
Total	$15 3/4$ billion years	

Many scholars believe that the book of Genesis was not intended to be a scientific description of the creation of the universe. Nevertheless, if the time period used in the book of Genesis was cosmic time as suggested by Professor Schroeder, there is amazing correlation between the time periods during which various developments occurred as theorized by scientists and as described in the book of Genesis.

Note that the total age of the universe in equivalent Earth time ($15 3/4$ billion years) based on Professor Schroeder's analysis is close to the current scientific estimate of the age of the universe. The analysis performed by Professor Schroeder, as described above, provides a way, based on accepted scientific principles, for both scientists and biblical scholars to agree on the age of the universe.

Even if you cannot accept the cosmic time/Earth time dichotomy described above, the description of creation in the book of Genesis appears to provide the opportunity for an explanation of the word "day" as meaning a period longer than 24 hours (as agreed to by William Jennings Bryan, the prosecuting attorney, when questioned as a witness in the 1925 Scopes trial). If you interpret the book of Genesis as saying that the sun was not created until the fourth day of creation, based on one interpretation of Genesis 1:16,[75] it would be difficult to have morning and evening on days one, two and three.[76] Thus days one, two and three could be "days" or "eras" that lasted billions of years. In any case, many scholars would agree that the main purpose of the book of Genesis is to describe God's relationship with human beings and not necessarily to be an accurate scientific description of the details of creation.

Consider also that the apostle Peter confirmed what the psalmists said: "But do not ignore this one fact, beloved, that with the Lord a day is as a thousand years, and a thousand years are as one day."[77] The apostle Peter is saying that, for an eternal being like God, time is not a limiting factor and that describing the length of time for an era might not be especially critical. This would seem to imply that the use of the word "day" in the Bible is not necessarily a definitive 24-hour period.

MODERN EVIDENCE CONSISTENT WITH THE THEORY OF EVOLUTION BY NATURAL SELECTION

The authors of *Ancestors* described two areas of modern scientific inquiry in which the evidence is consistent with the theory of

---

75. Genesis 1:16. "And God made the two great lights, the greater light to rule the day, and the lesser light to rule the night; He made the stars also." Revised Standard Version. Note that Professor Schroeder claims that many ancient biblical scholars interpret Genesis 1:16 as meaning that the atmosphere of the earth became transparent and the sun became visible from the surface of the earth.

76. Genesis 1:5, 1:8 and 1:13.

77. 2 Peter 3:8 and Psalm 90:4. Revised Standard Version.

evolution by natural selection. These include research in genetics and in the behavior patterns and traits of primates in the wild.

GENETICS AND EVOLUTION

DNA and RNA are special arrangements of nucleic acids in the center of every plant and animal cell. DNA, which makes up the genes and, along with RNA, regulates all of the characteristics of the organism, was discovered in the 1950s and was not known to Darwin. Biologists during the time of Darwin knew that traits were passed down from generation to generation, but they did not know the mechanism. With the discovery of DNA and RNA, scientists now have a way of explaining how traits can be passed down, how there can be genetic variation, and how genetic mutations can occur that result in the traits of the organism being modified.

The authors described in *Ancestors* how more than 99 percent of the active genes in the DNA of humans is identical to that of chimpanzees based on the sequences of the four amino acids that make up DNA, which are abbreviated as A, C, G and T.

> *When ACGT sequences that are mainly active genes are examined, a 99.6% identity is found between human and chimp. At the level of the working genes, only about 0.4% of the DNA of humans is different from the DNA of chimps.*[78]

The authors of *Ancestors* also claim that the genetic evidence indicates humans and chimps are the two closest primates:

> *On the basis of all the evidence, the closest relative of the human proves to be the chimp. The closest relative of the chimp*

---

78. Sagan (1992), pg. 277 quoting from M. Goodman, B.F. Koop, J. Czelusniak, D. H.A. Fitch, D.A. Tagle, and J.L. Slightom, "Molecular Phylogeny of the Family of Apes and Humans," *Genome 31* (1989), pp. 316–335; and Morris Goodman, private communication, 1992. Similar results are found from DNA hybridization studies: C.G. Sibley, J.A. Comstock, and J.E. Ahlquist, "DNA Hybridization Evidence of Hominoid Phylogeny: A Reanalysis of the Data," *Journal of Molecular Evolution 30* (1990) pp. 202–236.

is the human. Not orangs, but people. Us. Chimps and humans
are nearer kin than are chimps and gorillas or any other kinds
of ape not of the same species.[79]

The authors of *Ancestors* go on to caution, however, that even
this small genetic difference of 0.4 percent can cause significant
differences.

> . . . *even a 0.4% difference could, for all we know, imply*
> *profound differences in certain characteristics [due to the poten-*
> *tial effect on enzymes, which have a powerful leverage in the*
> *body, and due to genetic promoters and enhancers].*[80]

The genetic evidence presented in *Ancestors* (that humans
and other primates are closely related) is impressive. As stated
above, more than 99 percent of the active genetic coding of
humans and chimpanzees is identical. This provides scientific
evidence for what we might have guessed from a visual comparison of these two groups of primates. As described in *Ancestors*,
in the 1860s, Thomas Huxley pointed out the physical similarities between humans and chimpanzees. Humans and chimpanzees have very similar skeletal structures: a skull attached at the
top of a spine with two forward-looking eyes, the same number
of teeth, two ears with each ear on the side of the head, a nose
and a mouth, a hand with an opposable thumb and four fingers
at the end of two similar arms, a foot and five toes at the end
of two similar legs attached to a pelvis, and similar rib cages.
They also have similar brains, hearts, lungs, and other bodily
organs. They have similar bodily functions for nourishment,
breathing, and reproduction. They interact with their environment in similar ways through the senses of stereoscopic sight,
hearing, smelling, tasting, and touching.

The authors of *Ancestors* believe that the similarities in the
physical structure and genetic makeup of chimps and humans

---

79. Sagan (1992), pg. 277.
80. Sagan (1992), pg. 279.

provide overwhelming evidence that humans and chimps evolved from a common ancestor. Nevertheless, there are other potential explanations for these similarities that do not rely totally on the theory of evolution. There are several ways these similarities between humans and chimpanzees could be explained, including the following three scenarios:

- **Scenario 1**—The genetic code and physical characteristics of humans and chimpanzees are similar because they evolved from a common ancestor through natural selection (the theory of evolution).
- **Scenario 2**—The genetic code and physical characteristics of humans and chimpanzees are similar because they were made by the same designer or Creator in a relatively short period of time (the theory of special creation, a.k.a. "creation science," and the theory of intelligent design).
- **Scenario 3**—The genetic code and physical characteristics of humans and chimpanzees are similar because they were made by the same designer or Creator who used gamma or cosmic rays to affect genetic mutations and thereby influence how they would evolve from a common ancestor (a combination of the theory of evolution and the theory of intelligent design).

By identifying the above three possibilities, I do not mean to imply that the evidence for evolution should be discarded. Nor do I wish to imply that scientific evidence supports all three scenarios. I mainly want to point out that although the close genetic similarity between humans and chimps is *consistent* with evolution, it does not *require* evolution by natural selection as an explanation. A "scientific" explanation is an explanation that is supported by observable evidence. Based on currently available evidence, evolution by natural selection is the best "scientific" explanation. Some of the available evidence is described earlier in this chapter under the heading "Evidence Supporting the Theory of Evolution." Also, we would expect to find close genetic similarity if species developed according to the theory of evolution.

Note that the evidence for Scenario 1 and Scenario 3 could be very similar, making it difficult if not impossible to demonstrate whether or not a Creator was involved. As described below, there is scientific evidence that genetic mutations are caused by radiation.[81] Thus, it might be possible to discover evidence which shows how radiation has caused genetic mutations that resulted in humans and chimpanzees evolving from a common ancestor. Nevertheless, it might ultimately be impossible to use the evidence and scientific methods to determine whether Scenario 1 or Scenario 3 is the correct explanation. For example, it would be difficult if not impossible to acquire evidence that would indicate whether a certain genetic mutation was caused by random radiation and random mutation (Scenario 1) or radiation directed by an intelligent designer (Scenario 3). It would likewise be difficult to acquire evidence that would indicate whether or not an intelligent designer has been directly involved in the evolution of human bodies and chimpanzee bodies from a common ancestor.

Scenario 2 raises other questions. If all species were created by an omnipotent Creator at one point in time, why should there be such close genetic similarities between species? Why would there be greater similarity between animals on islands and the nearby mainland than between the animals on separate islands that have similar environments? There would be scientific evidence for Scenario 2 if it could be shown that evolution has occurred more rapidly than could be explained by natural rates of mutation and natural selection. That is the reason many people have been interested in the theory of punctuated equilibrium which initially appeared to provide evidence that species arose quickly during the Cambrian "explosion." However, as described above, Professor Miller explains that the observable rate of change based on natural mutation rates is many times faster than needed to explain the development of species during the Cambrian period.

---

81. Note also that since the development of the atomic bomb and atomic energy in the 1940s there have been numerous science fiction movies based on genetic mutations caused by exposure to nuclear radiation.

Even if evolution (without the influence of an intelligent designer) is the way the various species developed on earth, this does not preclude there being an intelligent designer involved in the development of life. As described above, there are obstacles to the initial development of DNA and RNA by a completely natural mechanism. Thus, there is a potential role for an intelligent designer at least in forming the first single-celled creatures with DNA and RNA. In addition, the seeming random, directionless nature of evolution does not provide strong evidence against the existence of a divine Creator. As an eternal being, a divine Creator would have an unlimited amount of time. Such a Creator could let evolution take its random course and wait millions and billions of years until just the right circumstances occur for conveying a soul on just the right type of creature. If necessary, a divine Creator could have created hundreds of billions of universes, each with hundreds of billions of galaxies, each with hundreds of billions of stars until just the right star with the right planet (this one) had life which evolved in just the right way. This could be known as a "many universes" theory based on supernatural requirements.[82]

Radiation and Mutation

Many scientists—who accept evolution by natural selection as the mechanism under which the various species developed—believe that species have changed and evolved into other species by genetic mutation. One cause of genetic mutation is likely the background radiation that is continually present on the earth. As explained by Erwin Schrödinger (the father of the quantum wave equation)[83] in his 1944 book *What Is Life?*:

> *The percentage of mutations in the offspring, the so-called mutation rate, can be increased to a high multiple of the small natural mutation rate by irradiating the parents with X-*

---

82. Scientists also have several "many universes" theories.

83. See the chapter "Quantum Mechanics."

rays or gamma rays.[84] The mutations produced in this way differ in no way (except by being more numerous) from those occurring spontaneously, and one has the impression that every "natural" mutation can also be induced by X-rays. This implies natural mutations might be caused by natural background radiation.[85]

Consider then that under Scenario 3 (described above) an intelligent designer, who is interested in developing various forms of life on earth, could direct radiation to affect the mutation of DNA and RNA in precisely the way needed to produce the various species that have appeared on the earth. If the intelligent designer is the same being that created the universe, it is not difficult to imagine that this Creator/designer could also work "behind the scenes" to pinpoint the trajectory of a high-energy electromagnetic ray, such as an X-ray or a gamma ray, to bombard just the right genes and cause the desired mutations. Note that X-rays and gamma rays are electromagnetic radiation, which is physically the same type of radiation that, in everyday language, we call "light." The X-rays and the gamma rays have a higher frequency than visible light, which can be detected by our eyes. A Creator who can create light (such as the light present in the big bang) would also be able to create X-rays and gamma rays adjacent to the appropriate gene to cause it to mutate. This might have been the first micro-laser surgery or genetic engineering.

SOULS

So where do supernatural souls fit in the above three scenarios? The answer is that they are compatible with and necessary in each of the three scenarios. Regardless of how the human body might have developed—either at the hand of an intelligent designer or by natural selection over hundreds of thousands or

---

84. X-rays and gamma rays are high-energy electromagnetic waves (light) which oscillate at a very high frequency.

85. Schrödinger (1944), pg. 42.

millions of years—supernatural souls are needed to explain the ability of humans to make free choices.

Under Scenario 1, God might have infused a male and a female of a primate species (or a whole group of the species) with supernatural souls after that species had evolved through natural selection to the point when the right characteristics of human anatomy and intelligence were present. Under Scenario 2, God might have infused a soul into the first male and female humans when they were created. Scenario 3 would be the same as Scenario 1 except that the bodily evolution of the species would be due to divine intervention rather than random mutation. For all of the scenarios, God would need to infuse a soul into each and every descendant of the first two humans (or the descendants of the group). Souls are not a part of the body; they are not subject to the laws of science or genetics. Souls cannot be inherited like bodily genetic traits.

Recognizing that human free will cannot be explained by natural phenomena that follow the laws of chemistry and physics is something scientists as well as religious and spiritual leaders can agree on while the evidence concerning the origins of the bodies of humans is further discovered and examined. Recognizing that a human being cannot be explained completely as a material creature can unify us while we continue figuring out the exact mechanism of how and why our bodies look and act like they do. A question for scientists: Is the human body any less magnificent if there is also a supernatural soul? A question for the religious and spiritual: Are humans any more or less dignified depending on whether our bodies were formed out of mud (as per a literal interpretation the book of Genesis) or whether they evolved from other animals? In either case, it is clear that our bodies (including our brains) begin as tiny one-celled organisms and are ultimately made up of natural matter and elements.

The presence of supernatural souls in humans would certainly give God plenty to do. It would be a monumental job to supply souls for the thousands of new humans who are born every day and millions who are born every year, not to mention the billions of humans who have lived since the dawn of history. Rather than having a God for whom there "is nothing left to do," as Carl Sagan would have us believe, we now have a God that is doing (as far as humans are concerned) the most important work in the universe, that is, creating human souls.

While it is important to continue on our search to understand the physical origin of humans, it is likewise important to keep in mind that a divine being could use any of the scenarios described above, including evolution, to produce a race of creatures to which He bestows a supernatural soul.

There are many unresolved issues in all of the three scenarios described above; and they are likely to remain unresolved for years, if not decades or even hundreds of years, as we continue to gather more and more evidence and to develop more sophisticated tools to investigate the evidence. As described above, even with more evidence, it might ultimately be impossible to determine under which scenario the human body developed.

While we are waiting for the "final" evidence concerning evolution to be found, we have a choice: Are we going to accept the evidence that all natural phenomena interact according to the laws of physics and chemistry and that humans have free will? Are we going to accept the logical conclusion that free will cannot be explained as a natural phenomenon? Along with this is the logical conclusion that a supernatural soul makes us a special creature in this world. I hope that both the scientific and religious communities can support these concepts so that humans throughout the world can believe in the inherent spiritual worth of all human beings and act accordingly.

Similar Behavioral Characteristics

Relying on the work of such pioneers as Jane Goodall who has spent her adult life living with and recording the habits and actions of chimps in the wild, the authors of *Ancestors* were able to compare the behavioral characteristics of humans and chimpanzees. Many traits common to both groups can be documented as described below. The authors of *Ancestors* theorize that the existence of similar behaviors by both groups provides strong evidence that they evolved from a common primate ancestor:

> *We're not descended from chimps (or vice versa); so there's no necessary reason why any particular chimp trait need be shared by humans. But they're so closely related to us that we might reasonably guess that we share many of their hereditary predispositions—perhaps more effectively inhibited or redirected, but smoldering in us nevertheless. We're constrained by the rules that, through society, we impose on ourselves. But relax the rules, even hypothetically, and we can see what's been churning and fermenting inside us all along. Beneath the elegant varnish of law and civilization, of language and sensibility—remarkable accomplishments, to be sure—just how different from chimpanzees are we?[86]*

First of all, note that the authors of *Ancestors* recognize that humans have "rules . . . we impose on ourselves." These rules are not imposed on humans by evolution or by DNA/RNA but by free will. Since humans have free will, we are able to "relax the rules" as the authors propose. If humans were solely governed by evolution or DNA/RNA, we would not be able to either impose the rules or relax them. Even though we share many of the same behavior traits with chimps, we do not share free will. It is this very ability to choose whether or not to "impose rules" or "relax the rules" that is one of the most, if not the most, important difference between humans and chimps, as well as all other animals. Free will is such a basic part of human nature that it is often overlooked. The comment by the authors of *Ancestors* that we are no different from chimps if we relax the rules is equivalent to saying that we are no different from chimps except that we are *radically* different because we have free will.

---

86. Sagan (1992), pg. 313.

Table 4.2

Human Characteristics that Animals Supposedly Do Not Have	Countervailing Examples of Similar Animal Characteristics
Humans make tools.	Chimps make "tools" out of sticks to extract termites for food.
Humans are political.	Chimps have hierarchical societies.
Humans are ethical and moral.	Scientists have conducted an experiment in which a macaque would go hungry rather than press a button to receive food when it could see that pressing the button would also cause another macaque to experience pain.
Humans have self-awareness.	Chimps have been observed to recognize that markings on their image in a mirror are on their own bodies.
Humans have an elaborate culture.	Chimps also have a complex culture.
Humans have verbal speech.	Chimps can use verbal sounds to communicate.
Humans have private property.	Chimps hide food.
Humans have rational thought.	Chimps make rational choice of materials for homemaking.
Humans establish cities, states, and countries.	Chimps defend a group territory.
Humans have an intellect.	Chimps can interact in very complex situations.
Humans can understand.	It is difficult to demonstrate what chimps are thinking.
Humans cook their food.	Chimps do not have the dexterity to do this but have invented new methods to make food useful.
Humans think about the future.	Chimps hide food for future consumption.
Humans can think abstractly.	Difficult to demonstrate but chimps hiding food for future consumption seems to have an abstract quality about it.

After examining many supposed differences in the behavioral characteristics between humans and chimpanzees that have been proposed by scientists and philosophers over the years, the authors conclude that there are only differences in degree and not in substance. The preceding table provides a list of some of the "differences" between humans and animals that have been proposed historically and the countervailing examples given in *Ancestors*, although I do not necessarily agree all the countervailing examples validly refute the difference. Other examples of differences for which I am not aware of any valid countervailing examples are farming, cooking food, and developing the complex technological society in which humans live, all of which require a high level of conceptual thinking, planning, and understanding.[87]

In addition to the above refutation of the supposed differences between humans and animals, *Ancestors* presented a discussion of certain traits of chimps in the wild (revealed by the research of Goodall) that are similar to human traits. Some of these traits are listed below:

> *Hold grudges, nurse resentments, harbor thoughts of revenge, plan future courses of action, defend their children, raise orphaned infants by siblings, experience prolonged grief at the loss of a loved one, suffer from human diseases, turn gray, get wrinkles, lose teeth, lose hair, get drunk, understand words, recognize themselves in a mirror, get cranky and irritable when they're weaned, form friendships, share food, keep secrets, lie, steal, oppress the weak, protect the weak (even at the risk of danger to themselves), dominate others, submit to authority, pay respect to authority, challenge authority, operate within a hierarchy, exhibit a show of strength to maintain status, fight to maintain status, intervene to prevent conflict, strive for social advancement, accept their lot in life, socialize*

---

87. See the 2003 book *Modern Physics and Ancient Faith* by Stephen Barr for a more thorough discussion of the differences between humans and animals in mental ability.

with others, learn trails, recognize chimp voices, communicate about sex, show dominance, recognize hidden dangers, bury food supplies, beg for food, play, fight mock battles, (males) protect females and the young, commit infanticide, participate in cannibalism, eat meat, hunt for meat, (males) rape females, comfort the grieving, have temper tantrums, threaten violence, commit acts of violence, kill others of their own species, plan ambushes, cooperate in hunting or on patrol, exhibit male bonding, protect territory, exercise self-control, engage in deductive reasoning by examining the clues on a hunting trip, attack and kill strangers, celebrate after success, delight in acrobatics, bluff and intimidate others, act angry, try to annex territory, make mock sexual gestures of dominance, copulate with more than one partner, (males) reassure one another by touching each others' testicles (as the ancient Hebrews and Romans are said to have done upon concluding a treaty, or testifying before a tribunal. Indeed the root of "testify" and "testimony" is the Latin word, testis.), groom each other, form alliances, change alliances, act as peacemakers or mediators, act more community-minded in crowded conditions, act jealous, (males) attack the children of sexually unresponsive females, (females) nurse and nurture the young, abhor incest, cherish their parents, males on average are larger and stronger than females, (males) react aggressively to high testosterone levels, learn new activities from their peers, invent new methods to make food useful.[88]

It is clear from the above list that many of the behavioral traits of chimps are similar to the behavioral traits of humans. However, as discussed above, humans are different from chimpanzees in at least one important way: free will. As described in the chapter "Free Will Test," the free will test provides evidence that humans have free will. However, there is no evidence that chimps have free will or are able to make free choices. Although

---

88. These traits are described on pages 282–315 in *Ancestors*.

animal intelligence has been well documented, there is a qualitative difference between intelligence and free will. Intelligence enables an organism to identify various situations and conditions in its environment and define the best reaction to those situations and conditions. Free will, on the other hand, enables humans to choose whether or not to pursue whatever is determined to be the "best reaction."

Ancestors addresses the question of free will in humans and animals. Regarding humans, *Ancestors* suggests that at times humans appear to be in control of their actions and at times they do not. I think that close reflection will reveal that unless there is some type of physical or mental impairment, humans can choose to be "in control" of their actions or they can choose to follow their feelings. Following feelings may appear to be "out of control" but it is a choice nonetheless. In the final analysis, however, humans are either capable of free will or they are not, regardless of the fact that they do not appear to be in control at certain times. Apparently the authors of *Ancestors* believe that humans do have free will as they claim that "we are free:"

> *We are free to posit, if we wish, that God is responsible for the laws of Nature, and that the divine will is worked through secondary causes. In biology those causes would have to include mutation and natural selection.* [89]

Apparently, Professor Sagan did not notice that Professor Hawking claimed in *A Brief History of Time* that free will is an illusion, which contradicts the above statement by Professor Sagan that "we are free." If Professor Sagan had noticed this radical claim by Professor Hawking, he might have changed his opinion as expressed in the forward of *A Brief History of Time*. [90]

---

89. Sagan (1992), pg. 63.

90. See the discussion about Professor Sagan's comment earlier in this chapter.

Regarding animals, the authors of *Ancestors* wonder whether animals might have free will like humans but they do not offer any evidence either for or against it:

> *"[M]an differs from irrational creatures in this, that he is master of his action," was a tenet of St. Thomas Aquinas in his **Summa Theologica**. But are we "masters" of our actions always and in all circumstances? Do other animals never exhibit "mastery"?*[91]

The authors of *Ancestors* relate the example given by Thomas Aquinas in his 1270 treatise *Summa Theologica* in which he discusses the situation of a stag choosing which direction to go at a fork in the road. Thomas Aquinas rejected the notion that a stag can freely choose because he reasoned that a stag does not have free will. This seems to be circular reasoning because it appears he simply assumed that a stag does not have free will. I agree with the conclusion that a stag (or any animal) does not have a free will, but I base this on there not yet being any evidence that a stag or any animal (other than humans) can make decisions that are not based on a reaction to their instincts or genetic coding. For example, there are a number of factors that could affect a stag's decision at a fork in the road: the need to return to familiar territory, the need for water or food, or the need to avoid predators. The hesitation at the fork in the road could be due to the time it takes for the stag's brain to process the information and come to a conclusion. The decision could be based on any one or more of the above factors and how the stag's brain has been programmed to respond. There is no evidence, though, that the stag can make a free choice to override how it has been programmed to respond.

It may be that humans and other animals have similar "thoughts" when they are hungry, thirsty, and tired. Certainly their responses are very similar: they eat, drink, and sleep. But

---

91. Sagan (1992), pg. 369.

humans have the ability to interject choice into every step of their lives: what they have to eat and drink, what they do each day, where they go, whom they see, whether they work or play, what they say, when they go to bed, etc. In humans, instinct and free will are not contradictory, they are complementary. Human bodies give strong signals that are needed to preserve life: when to breathe, when to pump blood into our arteries, when to eat, when to drink, when to sleep, etc. To some of these signals, the bodies respond automatically and without the need to make a conscious decision: breathing, the pumping of blood, the operation of the digestive system. Nevertheless, humans can consciously intervene and "hold their breath." There have even been reports of the ability of some people to consciously affect their heart rate.[92]

Ancestors hypothesizes that humans are not conceptually different from chimps since they have the same thoughts and instincts as chimps. In his 1961 best seller, *African Genesis*, Robert Ardrey[93] proposed a similar premise: that humans have "killer instinct" because they evolved in Africa from flesh-eating killer apes. He also discussed other animal instincts in humans: territorial protection and hierarchical societies that have since been confirmed in great detail by the research of Jane Goodall.

It is old news, however, that humans have "animal" instincts. Most religions have taught for thousands of years that humans have natural instincts. However, they also teach that humans must often use free will to choose to reject where the natural instincts might lead: to reject violence, killing, raping, stealing (including stealing territory in wars), jealousy, hating, lording over others, etc. The fact that humans can choose to follow their instincts and kill other humans is as old as the human race. The fact that these feelings are "natural" or are due to having "animal instincts," however, does not mean that it is morally right to follow them. The book of Genesis, which was written thousands of years ago,

---

92. Chopra (1993), pg. 13.

93. Mr. Ardrey is also author of *The Territorial Imperative*.

reports the story of how Cain, one of the first humans, killed his brother Abel due to jealousy. The ability of humans (unlike other animals) to choose to reject their instincts and live moral lives is also well known. Two thousand years ago the Apostle Paul wrote "For the sinful nature desires what is contrary to the Spirit, and the Spirit what is contrary to the sinful nature."[94]

DOES EVOLUTION JUSTIFY ATHEISM OR THE BELIEF THAT HUMANS DO NOT HAVE SOULS?

I would suggest that in and of itself, the theory of evolution does not actually address whether or not there is a God or human souls. The theory of evolution, as with all scientific theories, relies on the operation of natural forces and makes no overt claims about the spiritual or supernatural world. Evolution does not have a need for a God or a supernatural human soul but neither does it preclude the existence of either. As explained in *Ancestors*:

> Evolution in no way **implies** atheism, although it is con-
> sistent *with atheism.*[95] *[Emphasis in the original.]*

I agree that evolution does not imply atheism. But knowing that evolution cannot explain free will in humans, is atheism consistent with the required supernatural nature of humans needed to explain free will? There now has to be a dual explanation for the origin of humans. In addition to the explanation for the origin of the human body, it is logical to conclude that we need a supernatural soul to explain human free will. It is also logical to conclude that we need a supernatural God to explain the creation of souls. Even if it were assumed that evolution is an accurate description of how natural life on earth developed, we would nevertheless need a supernatural God to infuse souls into creatures that had evolved from other life forms over time. Thus, I believe that atheism can no longer be logically supported due to the evidence

---

94. Letter of the Apostle Paul to the Galatians 5:17.
95. Sagan (1992), pg. 66.

for the existence of human free will and the evidence that the interaction of all natural phenomena is subject to laws, thereby requiring a supernatural soul to explain free will.

Humans Are Stewards of the Environment

Because of free will, humans can choose whether or not to follow their instincts. They are not just another animal interacting with the environment based on their instincts. Because of free will, humans can choose how they will affect the environment. With this freedom to choose comes the responsibility to be good stewards of the natural resources that have been given to humans. Because of this freedom to choose, it is not appropriate to take the approach that humans should "follow their instincts" and let evolution decide whether or not human actions are beneficial and whether or not humans will survive. In my opinion, humans must strive to use wisdom to protect the natural environment (including the air, the water, the soil, and all plants and animals) for current and future generations in balance with providing economic activity that enriches the impoverished people of the world. As with all areas of life, the right course of action can only be determined by engaging in thoughtful study, listening, and considering many viewpoints.[96]

Evolution Versus Devolution

Occasionally, there is controversy as to whether or not science and religion provide two irreconcilable views of human nature:
- Some people think evolution describes humans as a higher life form that has evolved over the ages in a gradual "improvement" over lower life forms.
- The image of humans presented in the book of Genesis is one of a fallen creature.

96. I also believe, based on the supernatural nature of humans, that prayer is helpful in reaching the right decision.

In my opinion, the apparent conflict contained in these views is easily reconciled. The first description of humans is a description of the material body. It is based on the assumption that intelligence represents a higher form of animal existence than other animal traits. Although humans are most likely the most intelligent animals on earth, from an evolutionary standpoint no one trait represents a "higher" life form unless that trait helps the animal to survive and reproduce. As long as intelligence enables humans to survive and multiply, humans will be considered to be a "successful" species.

Even if we accept that the human brain represents the pinnacle of animal evolution, the story of the fall of humans from grace contained in the book of Genesis is not incompatible with the superior intelligence of humans. The story in Genesis is concerned with the state of the supernatural soul of humans and not the body or the brain. The story in Genesis is about how humans can use their free will to reject the laws of God. Humans can use their free will along with their superior intelligence to either live in harmony with God or in a state of rebellion against God. There is no inherent incompatibility between superior intelligence (the "scientific" view) and free will (the "religious" view).

ADAM AND EVE

There are often discussions as to whether Adam and Eve were actual individuals or whether they simply represent the group of primates which could first be recognized as human beings. From the perspective of evolution, a case could be made for either interpretation. According to the theory of evolution, a group of primates would have had to become isolated from the rest of their original group, presumably by some type of geographic barrier or perhaps by migrating to another continent, for hundreds of thousands of years or longer. At some point, the separated group would have evolved enough to become a separate species that could no longer interbreed with the original group of primates from which it was separated. At that point (or later), the separated primates could re-enter the area inhabited by the original group of primates and continue to evolve without mixing genetically with the original group. At the same time (or later) the separated group would have evolved enough in terms of intelligence and other characteristics to be called "human." Under this scenario, the separated group that becomes human would be representative of Adam and Eve.

But there is also another possible evolutionary scenario. The separated group could remain separated from the original group until all of the intellectual and other characteristics of humans had evolved. Then, due to whatever barrier had originally separated the two groups, it is possible that only one couple from this separated group was able to rejoin the original group. It is also possible then that the rest of the separated group was never able to rejoin the main group and eventually died out. In this scenario, the human couple that was able to return from across the barrier would be the actual individuals Adam and Eve.

It is also important to remember that regardless of which scenario describes how the human body came to be, humans also have a supernatural soul. This means that at some point God infused a supernatural soul into humans. This further implies that there was a definite time when humans came into existence on the earth. This evidence is not consistent with the theory that humans changed slowly and imperceptibly over time, making it impossible to define exactly when human existence on Earth began and making it impossible to have an identifiable Adam and Eve. Even though a primate body might have evolved slowly, the point at which God gave humans a supernatural soul defined the point at which humans came into existence on the earth.

A supernatural soul could have been infused into a single couple (that either evolved or was created in a special act) or supernatural souls could have been infused into a whole group of primates that had the intellectual and other characteristics of humans. Thus it is possible, in my opinion, both from an evolutionary perspective and from a spiritual perspective that Adam and Eve were actual individuals.

It is interesting that in the 1980s and 1990s scientists became aware of evidence both that all humans have one common maternal ancestor and that all humans have one common paternal ancestor. The evidence for the common maternal ancestor is based on the fact that there is a type of DNA in the cells of animals that is not due to the combination of DNA from the male and female of the animal. Rather, this DNA is passed down from generation to generation only from the mother. It is called mitochondrial DNA ("mtDNA") and it floats around in the cytoplasmic fluid that surrounds the nucleus of the cell. Since the mtDNA does not join with DNA from the paternal parent, it changes very little from generation to generation. The mtDNA, however, is still subject to occasional mutation from background radiation. Some scientists have identified the different mutations in mtDNA by examining blood samples of living people from different backgrounds all over the world.

By estimating the average mutation rate for mtDNA, they have projected that all of the samples come from the same common female ancestor approximately 200,000 years ago. As described in the 1990 book *Searching for Eve* by Michael Brown, the abstract for the paper entitled "Mitochondrial DNA and Human Evolution," which was published in the January 1, 1987, issue of *Nature* magazine, read:

> *Mitochondrial DNAs from 147 people, drawn from five geographic populations, have been analyzed by restriction mapping. All these mitochondrial DNAs stem from one woman who is postulated to have lived about 200,000 years ago ... .*[97]

Note that the estimate of how long ago this "one woman" lived is based on the relative variation in mtDNA between the var-

---

97. Brown (1990), pg. 23.

ious geographic populations analyzed and an estimated mtDNA mutation rate. If the estimated mutation rate is wrong, the time that the woman lived could have been sooner or later. The estimate of when this "one woman" lived is given as a probable range between 140,000 and 290,000 years ago, with 200,000 years as the approximate average.[98]

It is also important to note that this data does not necessarily mean that this "one woman" was the first human. It mainly means that the other female descendents of the maternal grandmothers of this "one woman," those that were in other branches of the family, did not leave female offspring whose descendants survived to the present time. On the other hand, this "one woman" who lived approximately 200,000 years ago might have been the first human, if she was the first primate to have a soul, free will, and the ability to make free choices.

A similar search for "Adam" was based on an analysis of the "Y" chromosome which is uniquely a male genetic trait. As reported in the November 4, 2000, issue of the journal *Science News*, pg. 295, geneticist Peter Underhill of Stanford University and his colleagues probed alterations in the DNA sequence of the Y chromosome of 1,062 men from throughout the world. By performing a statistical analysis of the variations, they "constructed a tree of branching evolutionary relationships for men from the different parts of the world. Men from eastern Africa fell into a genetic group at the root of the Y chromosome tree." According to the Underhill team, "the genetic data behind both Eve and Y guy support the theory that modern humans originated relatively recently in Africa and then spread elsewhere, replacing groups such as the Neanderthals." Based on an estimated mutation rate, they estimate that "Y guy" migrated from eastern Africa into Asia between 35,000 and 89,000 years ago. Due to potential variability in the mutation rate, these conclusions about the origin of "Adam and Eve" are very controversial.

---

98. Brown (1990), pg. 32.

With a higher mutation rate, the time at which the human species appeared and began migrating could have occurred much more recently. It is interesting to note that with only modest increases in the size of each generation, humans, starting with one couple, could have easily reached the current population levels of the world over a span of only 6,000 years, the approximate time provided for by a so-called "literal" reading of the Bible according to the calculation made by Bishop Ussher in the mid-1600s. To demonstrate this, let us assume that on average each generation is 10 percent larger than the preceding generation, a rather modest growth rate. Thus on average each group of five couples (10 people) would have 11 surviving offspring[99] or 2.2 surviving offspring per couple, and these surviving offspring would go on to have children of their own. If we then assume each generation is on average 25 years apart, it would only take 5,300 years to reach a population level of 1 billion, the level of the world population in about 1850. The following table illustrates the relatively low population growth rates, ranging from 0.20 percent per year

Table 4.3

Assumed Average Population Growth Rate per Year	Equivalent Average Population Growth Rate per Generation	Number of Generations Needed to Reach a Population Level of 1 Billion	Years Needed to Reach a Population Level of 1 Billion (25 Years per Generation)
0.20%	5%	410	10,300
0.23%	6%	345	8,600
0.27%	7%	300	7,400
0.31%	8%	260	6,500
0.35%	9%	235	5,800
0.38%	10%	210	5,300
0.42%	11%	195	4,800
0.45%	12%	175	4,400

99. Eleven is 10 percent larger than ten.

to 0.45 percent per year, needed for the 1 billion population level to be achieved over a period ranging from 4,400 years to 10,300 years and starting with just one human couple.

In comparison, the population of the earth increased from 1 billion in 1850 to 6 billion in the year 2000, a six-fold increase in only 150 years. This represents an average growth rate of 1.20 percent per year over the 150-year period and an average growth rate of 35 percent per generation over approximately six generations during that 150-year period.

As described above, the growth rates of 0.20 percent to 0.45 percent per year required to achieve a population increase from one couple to 1 billion people as shown in the above table are significantly lower than the 1.2 percent per year growth rate achieved during the period from 1850 to 2000. However, this recent population growth from 1 billion to 6 billion in 150 years is probably not indicative of the growth rate that was achievable in ancient times when nutrition, medicine, and hygiene were not as good as in present times. On the other hand, during the early years of the human race, the life-threatening diseases known in recent historical and current times might not yet have developed.

Note that the development of the human race from a single couple to the current population level within the period of thousands rather than millions of years is consistent with the belief held by many scientists that all of human civilization has developed over the thousands of years since the last ice age:

> *Moreover, the most remarkable expansion of human mental powers (as witnessed by the development of cooking, agriculture, art, and, in a word, civilization) has all happened even more recently, since the end of the last ice age, in a 10,000-year twinkling that is as good as instantaneous from the evolutionary perspective that measures trends in millions of years.*[100]

---

100. Dennett (1991), pg. 190.

MEMES SUPPLEMENT GENES

Richard Dawkins,[101] who has written significantly in support of the theory of evolution by natural selection,[102] realized that human society seems to have advanced much quicker than can be explained by natural selection.[103] In his 1976 book *The Selfish Gene,* he devised a new concept called 'memes' that is meant to help explain the quick changes that occur in human society. In essence, memes are the ideas that are passed down from one human generation to the next generation. Just as there are mutations in genes, so there are changes in the ideas on which society operates. Through natural selection, beneficial genes survive and are passed down to future generations, while nonbeneficial genes do not survive. Likewise, beneficial memes, or good ideas, are generally kept by society and are passed down while nonbeneficial memes are not passed down. As Professor Dawkins explains:

> *Examples of memes are tunes, ideas, catch-phrases, clothes fashions, ways of making pots or of building arches. Just as genes propagate themselves in the gene pool by leaping from body to body via sperms or eggs, so memes propagate themselves in the meme pool by leaping from brain to brain via a process which, in the broad sense, can be called imitation. If a scientist hears, or reads about, a good idea, he passes it on to his colleagues and students. He mentions it in his articles and his lectures. If the idea catches on, it can be said to propagate itself, spreading from brain to brain.*[104]

---

101. Richard Dawkins is Charles Simonyi professor of the public understanding of science at Oxford University.

102. See also, for example, his 1986 book *The Blind Watchmaker: Why the Evidence of Evolution Reveals a Universe Without Design.*

103. Note, however, that there are biologists that have developed theories to explain the societal and cultural developments of humans using the principles of evolution by natural selection.

104. Dawkins (1976) 1989 ed., pg. 192.

I disagree with the idea that an idea can propagate itself. Because human beings have free will, they are able to select which ideas or memes to keep and pass on and which ideas or memes to discard.[105] A scientist is able to decide which ideas or hypotheses are supported by scientific evidence and which are not. Genes might be passed down by natural selection but memes are passed down by human selection. For example, the concept (or meme) of an ether[106] was considered by scientists in the 1800s. However, the hypothesis that an ether does *not* exist is supported by experimental evidence. Thus, the concept that an ether exists did not die out because the people that believed it died without passing it on (this would be natural selection). Rather, the concept died out because it was rejected by scientists (this is human selection) based on experimental evidence. In fact, the abstract idea of an ether is still known by scientists but its existence is simply not accepted. The concept of an ether did not by itself just automatically jump from brain to brain.

Professor Dawkins considers both genes and memes to be selfish replicators that seek to have themselves duplicated. He recognizes however, that humans alone have the power to reject nature (genes) and nurture (memes) that "created" them:

> *We are built as gene machines and cultured as meme machines, but we have the power to turn against our creators. We, alone on earth, can rebel against the tyranny of the selfish replicators.*[107]

---

105. Professor Dawkins' idea that ideas just jump from brain to brain did not jump to my brain without being rejected. Neither will my idea (that ideas do not propagate themselves) propagate itself. Each person must choose whether or not to accept my idea.

106. An ether was hypothesized to be an invisible substance that pervaded space and provided a medium through which light waves could travel. Evidence subsequently indicated that the speed of light is constant in all frames of reference, which implies an ether does not exist.

107. Dawkins (1976) 1989 ed., pg. 201.

I consider the previous excerpt to be a very good description of nonrobotic, human free will. Professor Dawkins' claim that humans can rebel against their genes (nature) and memes (nurture) is in direct opposition to Professor Provine's claim, described at the beginning of this chapter, that nature and nurture are all there is and that humans do not have free will that can be explained as biological mechanisms. Professor Dawkins' claim is also a recognition that humans are radically different from all other creatures and is in direct opposition to the concept that the differences between humans and other species are not qualitative differences but are only differences in degree.

Professor Dawkins' claim that humans can rebel against nature (genes) and nurture (memes) was strongly opposed by materialists who do not believe in the spiritual world.[108] They realized that human free will leads directly to the requirement that there be a nonmaterial entity that gives humans this freedom. As described below, Professor Dawkins appears to also intuitively realize that if he openly admits that humans have free will, then he must also admit they have supernatural souls. When Professor Dawkins was criticized for referring to humans as lumbering robots, he responded:

> *Part of the problem lies with the popular, but erroneous, associations of the word "robot." We are in the golden age of electronics, and robots are no longer rigidly inflexible morons but are capable of learning, intelligence, and creativity. Ironically, even as long ago as 1920 when Karel Capek coined the word, "robots" were mechanical beings that ended up with human feelings, like falling in love. People who think that robots are by definition more "deterministic" than human beings are muddled (unless they are religious, in which case they might consistently hold that humans have some divine gift of free will denied to mere machines). If, like most of the critics of my*

---

108. Dawkins (1976) 1989 ed., pg. 331.

"lumbering robot" passage, you are not religious, then face up to the following question: What on earth do you think you are, if not a robot, albeit a very complicated one? I have discussed all this in The Extended Phenotype, *pp. 15–17.*[109]

I will not comment on who I think has the muddled thinking. However, the portrayal in science fiction stories of robots that have free will and can love does not mean that such robots can in fact be made. Also, it is not clear from the above excerpt whether Professor Dawkins thinks that both humans and robots have free will or that both do not have free will. It appears that he rejects the idea of free will. His question to the "nonreligious" however is well put: Do humans have free will or are they just complex robots? His discussion in his 1982 book *The Extended Phenotype* does not resolve the issue.

> *Once again, of course, philosophers may debate the ultimate determinacy of computers programmed to behave in artificially intelligent ways, but if we are going to get into that level of philosophy, many would apply the same arguments to human intelligence (Turing 1950). What is a brain, they would ask, but a computer, and what is education but a form of programming? It is very hard to give a nonsupernatural account of the human brain and human emotions, feelings and apparent free will, without regarding the human brain as, in some sense, the equivalent of a programmed, cybernetic machine.*[110]

His reference to "apparent" free will appears to mean that he rejects the concept that humans have free will. Nevertheless, I think that his comment that humans can rebel against the tyranny of genes and memes demonstrates he does believe humans have free will. It is interesting that the excerpt from *The Extended Phenotype* relegates a nonsupernatural human to being a "programmed, cybernetic machine." This is another example of a sci-

---

109. Dawkins (1976) 1989 ed., pp. 270–271.
110. Dawkins (1982), pg. 17.

entist appearing to reject free will when faced with choosing[111] between the option of recognizing free will and a supernatural soul to explain free will or of rejecting both free will and supernatural souls.

CHAPTER CONCLUSIONS

1. All biological mechanisms function using atoms, molecules, and other natural phenomena. All evidence to date indicates that atoms, molecules, and other natural phenomena always interact according to the laws of physics and chemistry, thus leaving no room for free will. As long as atoms, molecules, and other natural phenomena interact according to the laws of physics, free will cannot be explained as a biological mechanism. Thus, free will could not have developed through any natural processes (including the process of evolution by natural selection). These conclusions are based on logical arguments and scientific evidence. They should be discussed and debated in every class in which evolution is taught.

2. Scientists almost unanimously believe that there is substantial evidence that supports the occurrence of evolution to explain the development of the various species of plants and animals, but a very small minority of scientists do not.

3. The theory of evolution does not preclude a spiritual God from intervening in the activities of this world through spiritual means.

4. For genetic mutations that occur due to radiation, it would be difficult to determine whether the mutations are caused by naturally occurring radiation or by radia-

---

111. But again, if he does not have free will, then he cannot choose whether or not he believes in free will.

tion created by an almighty supernatural Being to affect the development of various species.

5. The theory of evolution purports to explain the origin of species. It does not purport to explain the origin of life.

6. Scientists have not yet developed a scientifically testable theory to explain the origin of life.

7. The limitation relative to the ability of natural selection to explain the origin of life should be described whenever the theory of evolution is taught.

8. The discovery of a natural explanation for the origin of life would not demonstrate that free will can be explained as a biological mechanism.

9. All known living organisms are regulated by DNA and/or RNA.

10. With very modest increases in the population level from generation to generation, the 1850 world population level of 1 billion people could have been reached over a period of several thousand years starting with one human couple.

11. Ideas (memes) do not just jump from brain to brain. Humans use free will to either accept certain ideas or reject them.

"But I admit to some discomfort in working all my life in a theoretical framework [of quantum physics] that no one fully understands."

—*Professor Steven Weinberg*

"It is likely true that 'no one understands Quantum Mechanics' . . . although it is equally true that in some wonderful way Quantum Mechanics understands the Universe."

—*Professor Eugene Hecht*

"There was a time when the newspapers said that only twelve men understood the theory of relativity. . . . On the other hand, I think I can safely say that nobody understands quantum mechanics."[1]

—*Richard Feynman*

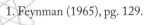
1. Feynman (1965), pg. 129.

Chapter Five

QUANTUM MECHANICS

W hat, you might ask, is a chapter on quantum mechanics (also known as quantum physics) doing in a book on the human soul? The main reason for its inclusion is to respond to the claims by several writers that free will might be explained by the quantum mechanical nature of the universe, rather than by a supernatural soul. Perhaps, they say, the source of free will can be found in the fuzziness of matter as described by the Heisenberg uncertainty principle or in the wave-particle duality of matter or in the collapsing of the wave function. Perhaps it is evident in the interaction of a human experimenter with a quantum mechanical experiment when he or she is choosing which aspect of nature to investigate. What is the nature of these concepts and can free will in fact be explained by quantum mechanics?

As described below, it is logical to conclude that matter alone (even matter that follows the laws of quantum mechanics) cannot be the source of free will. On the contrary, as described in the chapter "The Soul-Brain Interface," quantum mechanics provides a very reasonable explanation for the *mechanism* that allows for the interaction of a supernatural soul with a material brain. But quantum mechanics, in and of itself, does not provide an explanation of the *source* of free will.

To discuss why I think that it is logical to conclude that the source of free will cannot be a natural phenomenon that follows the laws of quantum mechanics, I will first explain some of the basic concepts of quantum mechanics. I assure the reader that I will focus only on the concepts of quantum mechanics and not the associated mathematical equations. While the concepts of quantum mechanics often seem counterintuitive, with a little patience they can be learned. For this discussion, most of the concepts will not be much different than the concepts of probability. An example of probability is the likelihood of getting heads or tails when we flip a coin. For this coin example the probability is 50 percent that we will get heads and 50 percent we will get tails.

Be aware, however, that, as described by several physicists, "no one really understands quantum mechanics." There are some unresolved conflicts in quantum mechanics that physicists have simply learned to accept.

DETERMINISTIC VERSUS PROBABILISTIC

In the 1600s, Isaac Newton developed laws of motion that were deterministic. This means that the mathematical equations which embody the laws give exact answers. With these equations, the exact velocity and position of a test item can be determined simultaneously, subject only to experimental accuracy. On the other hand, the equations which describe the laws of quantum mechanics produce probabilistic answers.[2] The locations of atomic particles are described in probabilistic terms. This means that we cannot simultaneously measure the exact position and velocity of a particle, but we can know the probability of finding it in certain areas. Usually, physicists describe probability shells which contain the particles but the exact locations of the particles are not defined. Note, however, that as described later in this chapter, some physicists suggest that it is more accurate to think of the underlying nature of matter and light as being bundles of waves rather than particles.

---

2. As described in this chapter, some physicists interpret the equations that govern quantum mechanics to be deterministic. Albert Einstein's famous observation about the probabilistic interpretation is that he did not believe that God plays dice with the universe.

Summary Conclusion

The results of this chapter can be summarized as follows: the probabilistic equations of quantum mechanics do not provide an explanation of the source of free will any better than the deterministic equations of Isaac Newton. Atomic particles (or waves) that are subject to the laws of probability do not have any more freedom to "do what they want" than macroscopic objects that are subject to Newton's laws of motion. Likewise, the fact that a human experimenter uses free will to choose which aspect of nature to investigate in a quantum mechanical experiment does not imply that free will is a natural phenomenon. On the contrary, since overwhelming experimental evidence indicates that all natural phenomena interact according to the laws of physics, this leads to the conclusion that the source of free will cannot be a natural phenomenon. This implies that the atoms and molecules that make up the brain of the experimenter cannot be the source of the free choice as to which aspect of nature to investigate. Neither can the atoms and molecules that are the subject of an experiment choose which aspect of nature they will display. Quantum mechanics might be the mechanism used to implement a free will decision via a human brain but atoms and molecules subject to quantum mechanical interactions cannot be the source of free will. Thus, it is logical to conclude that free will must come from outside the natural world.

Professor Steven Weinberg, who was awarded the 1979 Nobel Prize in physics, described in his 1992 book *Dreams of a Final Theory* that atoms do not have the freedom to behave "any way they want:"

> *Today, even though we cannot predict everything that chemists may observe, we believe that atoms behave the way they do in chemical reactions because the physical principles that govern the electrons and electric forces inside atoms leave no freedom for the atoms to behave in any other way.*[3]

---

3. Weinberg (1992), pp. 9–10.

Also, in the same 1992 book, Professor Weinberg explains:

> *Quantum mechanics is not deterministic in the same sense as Newtonian mechanics; Heisenberg's uncertainty principle warns that we cannot measure the position and velocity of a particle precisely at the same time, and, even if we make all of the measurements that are possible at one time, we can predict only probabilities about the results of experiments at any later time. Nevertheless we shall see that even in quantum mechanics there is still a sense in which the behavior of any physical system is completely determined by its initial conditions and the laws of nature.*[4]

From the above excerpts, we see that quantum mechanics does not give matter (including human brains which are made of matter) the freedom to have a "mind of its own." Whether the equations of quantum mechanics are considered to be probabilistic (the general understanding) or deterministic (the interpretation supported by some physicists), free will cannot be explained as the quantum mechanical interaction of natural phenomena. The following are typical quotations by physicists about the overwhelming scientific evidence that the natural world is governed by the laws of physics:

- *"Although scientists have never seen a violation of quantum mechanics in the laboratory (but have seen plenty of confirmations), the theory continually violates 'common sense.'"*[5]

- *". . . one can imagine a category of experiments that **refute well-accepted theories**, theories that have become part of the standard consensus of physics. **Under this category I can find no examples whatever in the past one hundred years.**"*[6] *[Emphasis in the original.]*

---

4. Weinberg (1992), pg. 37.

5. Kaku (1995), pg. 45.

6. Weinberg (1992), pp. 129–130

- *"... we now have in our possession laws that can describe correctly every experiment we have been able to invent."*[7]
- *"Quantum theory is our most successful physical theory ever. It can predict correct experimental results to an accuracy of several decimal points."*[8]
- *"Despite its baffling and strange version of reality, Quantum Mechanics has never once failed an experimental test. It is extremely reliable, though not transparently comprehensible. It is likely true that 'no one understands Quantum Mechanics,' although it is equally true that in some wonderful way Quantum Mechanics understands the Universe."*[9]
- *"Calculations using the mathematical formalism of quantum mechanics have been tested against countless laboratory measurements for almost a century, without a single failure. Quantum mechanics is often associated with 'uncertainty.' Nevertheless, it is capable of calculations to a high degree of precision."*[10]
- *"While the methods of quantum mechanics have proved their utility, no consensus exists on what quantum mechanics 'really means.' Some argue that the question itself is meaningless, that the mathematics speaks for itself."*[11]
- *"By 1928 or so, many of the mathematical formulas and rules of quantum mechanics had been put in place and, ever since, it has been used to make **the** most precise and successful numerical predictions in the history of science."*[12] *[Emphasis in the original.]*

7. Smolin(1997), pg. 16.

8. Zohar (1990), pg. 21.

9. Hecht (1998), pg. 5.

10. Stenger (1995), pg. 20.

11. Stenger (1995), pg. 21.

12. Greene (1999), pg. 87.

Likewise, from the very inception of quantum mechanics, physicists have recognized that human consciousness cannot be explained by the characteristics of quantum mechanics. In his 1971 book *Physics and Beyond: Encounters and Conversations,* Werner Heisenberg, the physicist who developed the uncertainty principle, recounted a 1931 conversation he had with Niels Bohr.[13] In their conversation they described how quantum mechanics cannot explain human consciousness:

> *"Another argument," I [Heisenberg] continued, "that is occasionally brought up in favor of an extension of quantum theory is the existence of human consciousness. There can be no doubt that 'consciousness' does not occur in physics and chemistry, and I cannot see how it could possibly result from quantum mechanics. Yet any science that deals with living organisms must needs cover the phenomenon of consciousness, because consciousness, too, is part of reality."*

> *"This argument," Niels [Bohr] said, "looks highly convincing at first sight. We can admittedly find nothing in physics or chemistry that has even a remote bearing on consciousness. Yet all of us know that there is such a thing as consciousness, simply because we have it ourselves. Hence consciousness must be part of nature, or, more generally, of reality, which means that, quite apart from the laws of physics and chemistry, as laid down in chemistry theory, we must also consider laws of quite a different kind."*[14]

Note that Professor Bohr recognized consciousness could be a part of reality without being a part of nature which is subject to the laws of chemistry and physics. Professor Bohr went on to express the hope that a better understanding of biology would provide an explanation for human consciousness. As described in

---

13. Niels Bohr developed the theory that electrons in atoms can only be found in certain discrete "orbitals" or energy states around the nucleus and that energy is released as light when electrons jump from one orbit or energy state to another.

14. Heisenberg (1971), pg. 114.

the chapter "Biology," the discovery of DNA which occurred 20 years later in the 1950s does not provide an explanation for free will which is a main aspect of human consciousness. Professors Heisenberg and Bohr also discussed how quantum mechanics cannot explain free will:

> [Heisenberg:] *"As you know, the fact that atomic processes cannot be fully determined is often used as an argument in favor of free will and divine intervention."*

> [Bohr:] *"I am convinced that this whole attitude is based on a simple misunderstanding, or rather on the confusion of questions, which, as far as I can see, impinge on distinct though complementary ways of looking at things. If we speak of free will, we refer to a situation in which we have to make decisions. This situation and the one in which we analyze the motives of our actions or even the one in which we study physiological processes, for instance the electrochemical processes in our brain, are mutually exclusive. In other words, they are complementary, so that the question whether natural laws determine events completely or only statistically has no direct bearing on the question of free will. Naturally, our different ways of looking at things must fit together in the long run, i.e., we must be able to recognize them as noncontradictory parts of the same reality, though we cannot yet tell precisely how."*[15]

As Professor Bohr stated, free will cannot be explained by physics regardless whether the laws of physics are deterministic or statistical (probabilistic).[16] Professor Heisenberg comments that some people look to the fact that atomic processes cannot be fully determined as evidence for free will and divine intervention. However, this is not a logical conclusion, and he does not discuss how he thinks the uncertainty of atomic processes could explain

---

15. Heisenberg (1971), pg. 91.

16. Professor Stephen Barr also discusses these concepts in his 2003 book *Modern Physics and Ancient Faith*.

free will and divine intervention. The uncertainty associated with measuring individual atomic particles does not mean that the atomic particles have "minds of their own."

As described later in this chapter, Erwin Schrödinger, another physicist who participated in the development of quantum mechanics in the 1920s, likewise concluded that a human being made only of matter cannot have free will.

The "Nonstandard" Muon

In February 2001, scientists performing a high-energy experiment reported that, for the first time, the motion of a muon, a subatomic particle detected in the experiment, did not follow the path described by the standard model of quantum mechanics.[17] The standard model is the currently accepted theory that explains the interactions of atomic and subatomic particles. Muons are subatomic particles that have an average life of 0.000002 seconds before they decay to form other particles. The motion of the muon, however, was consistent with another theory of physics known as string theory which is described later in this chapter. Of interest to the premise of this book is that there is no evidence that the muon behaved as though it had a "mind of its own."

A Short Summary of Physics

The next few pages contain a summary of some of the laws of physics that were known before the theory of quantum mechanics was formulated. This will provide a basis for understanding the revolutionary nature of quantum mechanics. The description of the "prequantum" laws is followed by a description of some of the main concepts that make up the quantum theory. Finally, there is a more detailed discussion of the claims that certain aspects of the quantum theory can be used to explain human free will.

---

17. *Minneapolis Star Tribune*, February 9, 2001, as reported by the *Boston Globe*.

The reader should be warned that it is not always possible to get a good mental image of certain concepts in quantum physics. One of these concepts is the wave-particle duality of nature, which is described below. The typical reaction when first exposed to the wave-particle duality concept is that the wave nature of matter and light is contradictory to the particle nature of matter and light. But rest assured that you are not alone in that response. All aspects of the wave and particle nature of matter and light cannot be contained in a single mental image. Physicists know that often the most that can be done is to learn how the physical world operates under certain conditions and how it operates differently under other conditions. Even though the two modes of operation might seem contradictory, the principles of physics require that both concepts be accepted as long as there is experimental evidence to support them. It is important to keep in mind that physics is basically concerned with an understanding of experimental results. It does not necessarily answer the question as to why something is the way it is.[18]

The World Before Quantum Mechanics

During the 1600s, scientists made great progress in discovering laws that describe the physical world. These laws enable scientists to predict how matter acts when subjected to defined forces. Sir Isaac Newton formulated the laws of motion that were very successful in describing the physical world, ranging from the interaction of small objects on earth to the motion of planets.

Mr. Newton used mathematical equations to describe the laws of physics in a way that is precise, clear, and unambiguous. Mathematics is used because it allows scientists to measure physical processes to test theories and, once a physical law has been established, to accurately predict how matter will act and interact.

---

18. For an in-depth discussion of the problems associated with devising a theory of physics that explains not only how things work but also why atomic particles have the characteristics they have, see *Life of the Cosmos* by Lee Smolin. Also see *The Elegant Universe* by Brian Greene.

Since Newton's laws of motion explain the universe in terms of a large, moving machine or mechanical process, they are known as the laws of "mechanics."[19] To distinguish these "classical" laws from the later "quantum" laws, the "classical" laws are often referred to as the laws of "classical mechanics" or "classical physics." The quantum laws of physics then became known as "quantum mechanics" or "quantum physics."

In the 1800s, the laws of energy and thermodynamics were added to the classical laws of physics. Remarkable advances in science and industry resulted from the better understanding of nature embodied by the classical laws of physics. The development of the steam engine provided a new energy source that greatly multiplied the productive capacity of humans and ushered in an era known as the industrial revolution. Tremendous bridges of steel and concrete were designed and constructed that enabled railroads to leap over natural barriers such as rivers. Steam-driven locomotives made possible the economical shipment of large quantities of raw material and finished products over long distances.

One of the most beneficial characteristics of classical mechanics is that the laws are very precise and deterministic. The laws of classical physics are in the form of mathematical equations, and as a scientist measures an experiment more precisely, the results more closely fit the established law. This deterministic nature of matter and forces is very beneficial in solving engineering problems. For example, in the design of a bridge, a given weight or force will deflect a beam or girder by a predictable amount. The designer can minimize cost by specifying bridge beams and girders that will support the amount of weight and force that is expected.

By the 1860s, James Clerk Maxwell had developed his well-known equations (wellknown, at least, to students of physics and electrical engineering) that describe the interaction between

---

19. Used in this sense, the word "mechanics" means mechanical things that operate like machines.

charged particles and electric and magnetic fields. Professor Maxwell's equations formed the conceptual and theoretical basis for an electrical revolution which ultimately witnessed the development of electrical power grids, electric motors, electric lights, telephone, telegraph, radio, television, telecommunications, microwaves, computers, and the thousands of electrical products that are now available.

Maxwell's equations were used to predict the existence and velocity of electromagnetic radiation,[20] which is also known by the more common term "light." To the physicist, the term "light" or "electromagnetic radiation" refers to a number of similar physical phenomena known as radio waves, television signals, microwaves, infrared light, ultraviolet light, X-rays, and gamma rays, as well as the visible light humans can detect with their eyes. The only difference between each of these types of light is the frequency of vibration.[21] For example, the frequency of visible light (approximately 1 billion megahertz) is between the frequency of infrared light (approximately 100 million megahertz) and the frequency of ultraviolet light (approximately 10 billion megahertz).

Maxwell's equations describe the relationship between magnetic and electric fields and charged electrical particles such as electrons and protons, which are discussed in more detail below. The use of Maxwell's equations demonstrates the power of mathematics and pure research. Maxwell developed his equations from what was known about electrical fields and magnetic fields, but before the scientific world knew that electromagnetic waves exist or that light is an electromagnetic wave. Nevertheless, Maxwell's equations were used to predict that the velocity of an electromag-

---

20. Electromagnetic radiation should not be confused with radioactivity, which is also often referred to as "radiation." Much of the radiation given off by radioactive material is not electromagnetic radiation although some radioactive materials give off gamma rays. The nonelectromagnetic type of radiation given off by radioactive material is made up of alpha and beta particles from the nucleus of the atom.

21. The frequency of vibration, or simply "frequency," is measured in cycles per second or hertz.

netic wave (if it existed) would be 300,000 kilometers per second (186,000 miles per second).[22] When, the speed of light was later measured to be 300,000 kilometers per second, it was postulated that light is an electromagnetic wave. This was subsequently confirmed by other experiments.

THE CLASSICAL LAWS FAIL

In the late 1800s, scientists began discovering physical phenomena that did not follow the laws of classical mechanics and electromagnetism. The laws of motion worked well when they were applied to physical objects encountered in everyday life. Problems with the laws of motion started appearing when scientists began investigating the nature of very small objects such as atoms.

All matter is made up of atoms, and the various kinds of atoms are called elements. A single atom of any element is the smallest particle which retains the physical and chemical properties of that element. For example, gold is an element and a single atom of gold still looks and acts like gold. Iron is an element and a single atom of iron looks and acts like iron. All in all there are 112 elements which make up all matter in the universe (as far as we know today).

During the late 1800s and early 1900s, physicists began to discover the parts of an atom. All atoms are made up of the same small building blocks called subatomic particles. All atoms have a center (a nucleus) that contains small particles called protons and neutrons. Each proton weighs about the same amount as a neutron but the proton has a positive electrical charge while the neutron does not have an electrical charge.

Around the nucleus of each atom are particles called electrons which have a negative electrical charge. An electron is much smaller than a proton or a neutron. Each proton and neutron

---

22. Velocity of an electromagnetic wave in free space.

is about 2,000 times heavier than an electron. Since electrically charged particles with opposite charges attract each other, each negatively charged electron is attracted to the positively charged protons in the center of the atom. Physicists initially theorized that electrons orbit around the nucleus of an atom similar to how the moon orbits around the Earth or how the Earth orbits around the sun. Sir Isaac Newton had theorized that the moon and the earth are held in their respective orbits by the force of gravity. Physicists likewise theorized that electrons are held in an orbit around the nucleus of an atom by the electrical attraction between the negatively charged electrons and the positively charged protons in the nucleus. As described below, this was a good start, but it was not the complete story. A conceptual sketch of a hydrogen atom (the simplest atom which has only one proton and one electron) is provided below. For this sketch, the electron is shown as a particle orbiting around the proton nucleus.

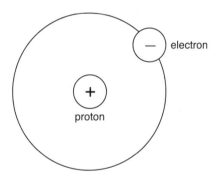

Figure 5.1

Hydrogen Atom: positively charged proton in the nucleus surrounded by negatively charged electron (orbiting as a particle).

(Simplified two-dimensional representation and not to scale.)

All of the various types of elements found in the universe are made up of subatomic particles: protons, neutrons, and electrons. The difference among elements is found in the number of protons that make up the nucleus of each element. For example, an atom of hydrogen (atomic number 1) has one proton in its nucleus. An atom of helium (atomic number 2) has two protons. An atom of lithium (atomic number 3) has three protons, etc. An atom that is electrically neutral has the same number of electrons as protons.

A molecule is made up of two or more atoms that are bound tightly together by sharing electrons. For example, a molecule of water is made up of two atoms of hydrogen (designated as "H_2") and one atom of oxygen (designated as "O") that are bound together by sharing electrons. A molecule of water is thus designated H_2O.

When an object circles around another object, classical physics theorizes there is a force acting on the orbiting object to make it change direction and circle around. According to classical physics, a force acting on an object causes it to accelerate. Thus, if an electron circles around the nucleus of an atom, by definition, the electron is accelerating because a force is acting on it to change its direction. An electron is a charged particle; it has a negative charge. Maxwell's equations predict that an electron that is accelerating as it circles the nucleus of an atom should continuously give off energy in the form of electromagnetic radiation:[23]

> *Classical electromagnetism predicts that charges will radiate energy when they are accelerated.*[24]

Scientists, however, have not detected radiation being continuously emitted from electrons orbiting around atoms. Also, if an

---

23. By definition, something is accelerating when its velocity is changing. Velocity is defined as the speed and direction of motion. An electron that is circling an atom is continuously changing its direction of motion and is thus accelerating. Under this classical model, the force acting on the electron to change its position is the attractive electrical force between the proton and the electron. This is analogous to the attractive gravitational force between the moon and the Earth which causes the moon to circle around the Earth. Like the moon, which weighs much less than the Earth, the electron weighs much less than the proton.

24. Halliday (1974), pg. 772.

electron is circling around an atom and radiating energy, it would ultimately lose all of its energy. This has not been observed.

Quantum Mechanics to the Rescue

Physicists are not happy when established laws and theories fail to predict how nature works. They quickly begin looking for a modified or alternative theory that will successfully explain all known experimental results. In the early years of the 1900s, physicists such as Max Planck, Niels Bohr, Werner Heisenberg, Louis de Brolie, Albert Einstein, and Erwin Schrödinger developed the quantum theory of matter and light which has been successful in describing how atomic particles interact. As explained above, these scientists discovered that the description of an electron orbiting around the nucleus of an atom was not entirely accurate. Rather, according to the quantum theory of physics, the electron is viewed as a "standing wave" that oscillates around the atom without radiating energy. The following sketch represents a typical standing wave in two dimensions.

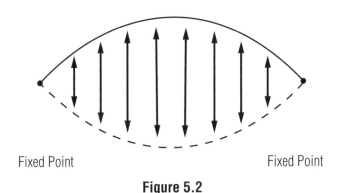

Fixed Point   Fixed Point

Figure 5.2

Simplified two-dimensional standing wave. The standing wave oscillates between the dashed line position and the solid line position.

Note that a standing wave has various points that are held fixed while the portion of the wave between the fixed points vibrates up and down. The standing wave in Figure 5.2 has a fixed point at each end, and the middle vibrates up and down. A well-known example of a standing wave similar to Figure 5.2 is the vibration of the string of a musical instrument when it is plucked. Typically, when the string of a musical instrument vibrates in its primary tone, it has two fixed points that do not move, one at each end of the string as shown in Figure 5.2. Between the two fixed points the string moves up and down, with the center of the string moving the most and the parts of the string nearer to the fixed points moving less. The movement of the string looks like a wave that is moving up and down, which is known as a "standing wave." A simplified two-dimensional representation of a standing wave of an electron around an atom is shown in the following sketch.

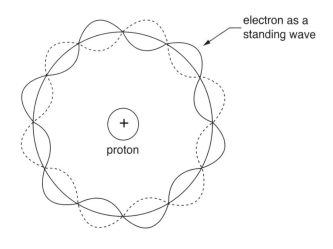

Figure 5.3

Hydrogen Atom: Positively charged proton in the nucleus surrounded by negatively charged electron represented by a standing wave. (Simplified two-dimensional representation and not to scale.)

In the above simplified sketch of an electron as a standing wave around the proton, the electron wave oscillates between the dashed line position and the solid line position, with the fixed points being stationary. Note that this sketch of the standing wave of an electron is shown only in two dimensions. An actual electron is a three-dimensional standing wave around an atom. In addition, this figure is a very simplified representation that is not intended to show what an electron "actually" looks like. Electrons oscillating around the nucleus of atom exhibit much more complex waveforms than shown here. Also, it is important to note that the wave does not represent the path of the electron around the atom. According to quantum physics, the electron *is* the standing wave which exists around the atom at all times.

Before you get too comfortable with the electron as a standing wave, remember that it also exhibits particle-like characteristics as well.[25] If you use an experiment to find an electron, you find it in one position. No one has ever directly measured the electron waveform around an atom. This inability to perform direct measurements of the waveform has been a source of great controversy since the beginning of quantum mechanics. Nevertheless, as evidenced by the quotations at the beginning of this chapter, quantum mechanics has been extremely successful in describing nature at the atomic level.

One of the principles of the quantum theory is that electrons in an atom cannot have just any level of energy. Rather, electrons can have only certain discrete levels of energy. For each of these discrete energy levels, the electron can be described as having a certain quantity of energy ("quantum" in Latin), which provides the name of the quantum theory. While scientists such as Louis De Brolie provided the theoretical concepts of the quantum theory, in 1926, Erwin Schrödinger developed a mathematical equation which predicts the discrete energy levels of electrons. In this equation, electrons are described as waves; it is known as the Schrödinger wave equation. The fact that electrons (and all matter) can behave like little tiny particles in some situations but can also behave like waves in other situations is known as the

---

25. For those interested, most physics text books describe some of the numerous experiments that can be done to show both the wave and the particle nature of subatomic particles.

wave-particle duality of nature. This will be described in more detail below.

Albert Einstein also contributed to quantum physics by theorizing that light travels through space in concentrated bundles of electromagnetic waves which are now called photons. Einstein's theory was that light travels as particles with each particle or photon having a specific quantity of energy. For each frequency of light there is a certain minimum quantity of energy into which light can be divided. This minimum quantity is the energy of one photon (one quantum particle of light). Although this particle nature of light appears to contradict the continuous-wave nature of light, we accept that light has both a particle and wave nature. Thus energy (both in the form of light and matter)[26] exhibits a wave-particle duality.

If you are confused at this point, don't feel alone. There is no one conceptual model that can completely explain the wave-particle duality. Probably the best model is to think of particles as being bundles of fields as described by Professor Weinberg. Note that, in general, these fields fluctuate as waves. Professor Weinberg explains the wave-particle duality in these words:

> . . . *all these particles are bundles of the energy, or quanta, of various sorts of fields. A field like an electric or magnetic field is a sort of stress in space, something like the various sorts of stress that are possible within a solid body, but a field is a stress in space itself. There is one type of field for each species of elementary particle; there is an electron field in the standard model[27], whose quanta are electrons; there is an electromagnetic field (consisting of electric and magnetic fields), whose quanta are the photons; there is no field for atomic nuclei, or*

---

26. Einstein's famous equation $E=MC^2$ means that energy is equivalent to the mass of matter. Matter converted to energy can show up in two forms: as the motion of matter (kinetic energy) or as light (pure energy without mass).

27. The standard model is the currently accepted theory that describes the interactions of atomic particles.

for the particles (known as protons and neutrons) of which the nuclei are composed, but there are fields for various types of particles called quarks,[28] out of which the proton and neutron are composed; and there are a few other fields I need not go into now. The equations of a field theory like the standard model, deal not with particles but with fields; the particles appear as manifestations of these fields.[29]

As described above, Professor Weinberg suggests we think of atomic particles as bundles of the energy of fields. To get a better understanding of a "field," try to put the north poles of two bar magnets together and feel how the bars are repelled by the magnetic field. Next, the Heisenberg uncertainty principle will be described and then there will be a discussion as to how the wave nature of matter and the Heisenberg uncertainty principle are used to try to explain free will.

THE HEISENBERG UNCERTAINTY PRINCIPLE

Another feature of the quantum theory some claim is related to the question of free will is the uncertainty principle developed by Werner Heisenberg (also known as the "Heisenberg uncertainty principle" or simply the "uncertainty principle"). The uncertainty principle states that it is impossible to measure with certainty the exact location of an atomic particle at the same time the velocity of the particle is measured. This makes it impossible to predict exactly where the particle will be in the future.

Here is where the deterministic model of the universe breaks down. Remember that in the world of classical mechanics, as described previously, it is possible to predict the future location of an object if the forces on the object are known along with the initial conditions (that is, the location and the velocity of the object

---

28. Quarks are even smaller subatomic particles that make up the protons and neutrons. However, quarks do not exist as separate particles. You cannot separate protons or neutrons into stand-alone quarks.

29. Weinberg (1992), pg. 25.

at the beginning of the experiment). The Heisenberg uncertainty principle, on the other hand, fits in well with the probabilistic model of the universe described by the Schrödinger wave equation.

An example of the deterministic model is a hockey puck on a flat, frictionless surface. Suppose it is moving five feet per second directly north. We can predict that in seven seconds the puck will be 35 feet from the starting point in a northerly direction. Likewise, if a steel ball is dropped and falls toward the earth, a scientist can use a mathematical equation to predict how far it will travel and how fast it will be going at any point in time after it is dropped.

The quantum theory embodied in Mr. Schrödinger's wave equation does not result in such a deterministic world. The wave equation does not predict the exact location of the electron. Rather, the wave equation is used to describe the probability of where an electron would be found in an experiment. For example, consider the smallest and simplest atom, the hydrogen atom which has a single negatively charged electron and a single positively charged proton as shown in the previous figure. According to the wave equation, the electron does not circle the proton in an orbit like

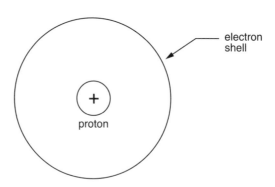

Figure 5.4

Hydrogen Atom: Positively charged proton in the nucleus surrounded by negatively charged electron represented by an electron shell. (Simplified two-dimensional representation and not to scale.)

the moon orbits the Earth. The most we know about the electron from the Schrödinger wave equation is that if we draw a spherical shell of a certain size around the proton, there is a certain probability of finding the electron (as a particle) inside the spherical shell.

If we make the spherical shell smaller to more precisely define the position of the electron, the probability of finding the electron (as a particle) in the spherical shell likewise decreases. This probabilistic nature of electrons is consistent with the uncertainty principle. If we knew the exact position and velocity of the electron at any point in time, we could predict exactly where it would be in the future. But due to the uncertainty principle, we cannot simultaneously measure the position and velocity of an electron. Thus, we cannot predict exactly where the electron will be and must talk in terms of probabilities.

THE PHYSICAL BASIS OF THE UNCERTAINTY PRINCIPLE

There is another approach that is useful in understanding the uncertainty principle. This has to do with the physical nature of experiments and measurement. Since the uncertainty principle is concerned with position and velocity of an atomic particle, it is instructive to understand how position and velocity are measured during an experiment. Often position and velocity are measured by taking photographs of the moving object at regular intervals and then noting the change in position of the object on each photograph.

For example, suppose a distance scale measured in meters is mounted in the test area and the object moves in front of the distance scale. Photographs of the object can then be taken at regular intervals, for example, one photograph every second. If the object is at the zero mark in the first photograph taken at time t = zero and is then at the five-meter mark in the second photograph taken at time t = one second, we can say that the average velocity of the object during the one-second interval was five meters per second. We can also say that the position of the object at time t = one second is at the five meter mark. To get more and more accurate measurements of the velocity and position of the object, the photographs could be taken at shorter and shorter intervals. Note that we can "see" the object by taking a picture of it because some of the light hits the object and bounces off of it. The energy of the light, however, is so small that it does not significantly affect the position or velocity of the object that is being measured.

Consider now performing a similar experiment to measure the position and velocity of an electron. An electron is much smaller than any macroscopic object we might want to measure. To get a clear, nonfuzzy image of the electron, it is necessary to take a "photograph" using light that has a wavelength that is smaller than the electron. As described above, the smaller the wavelength, the higher the frequency. And the higher the frequency, the higher the energy in each photon of light. The frequency and energy of the light needed to measure the position of an electron is so high that the light that hits the electron is strong enough to significantly affect the velocity of the electron. This makes it impossible to measure accurately the electron's velocity and position at the same time.

An analogy in the macroscopic world would be to try to determine the position and velocity of a ball by throwing 1,000 similar-sized balls at it at time t = zero and then noting the indentation of the thrown balls on a wall behind the target ball. The place on the wall without an indentation would be the place where the target ball was at time t = zero. Then, one second later, we would throw a second batch of 1,000 balls and see the indentations on the wall. (For this example, you have to assume it is possible to distinguish between the indentations of the first and second batch of balls thrown.) The problem with this method is that the thrown ball that hits the target ball would change the velocity and position of the target ball such that it would not be possible to determine how much of the target ball's velocity is due to its original velocity and how much is due to the other ball hitting it. Likewise, the velocity of the target ball would be changed when it is hit by the second batch of balls such that it would not have the speed and direction that would be indicated by comparing its first and second position.

THE UNCERTAINTY PRINCIPLE AND FREE WILL

There are some that claim the uncertainty embodied in the uncertainty principle is sufficient to explain the ability of humans to have a free will and make willful choices. Consider Professor Weinberg's statement:

> . . . *where the Newtonian physicist spoke of precise predictions, the quantum mechanician now offers only calculations of probabilities, thus seeming to make room again for human free will or divine intervention.*[30]

Professor Weinberg's statement is similar to Professor Heisenberg's comment to Neils Bohr excerpted at the beginning of this chapter. Neither Professor Heisenberg nor Professor Weinberg, however, has offered an explanation as to how "probabilities" can produce free will. In fact, the concept of free will being caused by probabilities directly contradicts Professor Weinberg's statement (quoted at the beginning of this chapter) that "even in quantum mechanics there is still a sense in which the behavior of any physical system is completely determined by its initial conditions and the laws of nature."[31]

Note that the indeterminate nature of matter as described by quantum physics is amazingly similar to the concept of an atomic "swerve" conceptualized in the 300s B.C. by the Greek philosopher Epicurus. Epicurus was trying to develop a philosophy that would free humans from the tyranny of the gods who controlled human lives by "fate." Epicurus was also not able to explain how

---

30. Weinberg (1992), pg. 77.
31. Weinberg (1992), pg. 37

this atomic "swerve" provided humans with free will other than to theorize that the swerve could somehow snap the bonds of fate that controlled humans. The Roman poet Titus Lucretius Carus included the Epicurean concept of the atomic "swerve" in his first century B.C. poem *De Rerum Natura (On the Nature of Things)* to attempt to explain the source of free will.

The argument for free will coming from probabilities, the uncertainty principle, and atomic "swerves," offered by others, goes something like this:

> *Due to the uncertainty evidenced by the uncertainty principle, the future movement and position of the atomic particles in the human brain are not precisely determined by the current movement and position of the particles. The human brain as it is "thinking" uses the nondeterministic nature of the atomic particles to make infinitesimally small changes in the movement and position of the particles of the brain to affect how the particles interact with one another. This, in turn, affects the brain neurons and synapses (which are made up of these atomic particles) and thus results in the signals that direct the rest of the body. Thus the human brain, by "thinking" affects how the atomic particles move and ultimately determines what signals are given to the rest of the body by the brain.*

The problem with the above explanation is that it does not explain what initiates a thought or a decision. It may be true that a decision can be implemented by affecting the nondeterministic position and momentum of particles in the brain but the source of the thought, decision, or free will choice must come from outside the brain, which is subject to the probabilistic rules of quantum physics. If "thinking" is only the interaction of atoms and molecules in the brain, then "thinking" itself is subject to the probabilistic rules of quantum physics. The uncertainty associated with the position and momentum of specific atoms and molecules does not provide an explanation, even conceptually, of free will. The interaction of particles under probabilistic rules does not provide free will any more than the roll of dice can produce free will.

Alternatively, as described below, it appears some, including Professor Weinberg, believe that free will is due to the inability to predict what the particles in the brain will do. However, free will requires an actual free choice and is not just the inability to predict a complex interaction. For example, a flag waving in the wind does not have free will just because it is too complex to predict what it will do.

An alternative but closely related explanation relies on the wave nature of matter and the collapsing of a wave function. As described below, the collapsing of a wave function cannot be the source of free will.

Collapsing the Wave Function

Some scientists have used the concept of "collapsing the wave function" to try to provide a physical explanation of free will or consciousness. Collapsing the wave function describes what happens when scientists measure the location of an atomic particle. When the position of a particle is measured, the particle exhibits its "particle" nature rather than its "wave" nature. For example, when the position of an electron around an atom is measured, the electron is found at a certain point rather than being found as a standing wave spread out around an atom. In this sense, the standing wave function "collapses" to a single point where the particle is found. As described below, some people believe that it requires a "conscious" being to collapse a wave function. In his 1996 book, *Empire of Light*, Professor Sidney Perkowitz describes the concept of collapsing the wave function as follows:

> . . . the meaning of the wave function has been elusive. In 1926, Max Born of the University of Gottingen . . . gave the interpretation we still use. The wave function is a map of potentiality, not a picture of reality. It gives the probability that an electron resides at this location or possesses that energy. The electron in a hydrogen atom is most likely to lie one-twentieth of a nanometer[32] from the nucleus, but its wave function

---

32. A nanometer is one-billionth of a meter. A meter is about 39 inches long.

also spreads into other locations where the electron may also exist. The actual location of the electron, however, is found only by measurement.

A wave function is like a list of the playing cards in a deck. The list yields statistical probabilities—for instance, that over many drawings the ace of spades appears one fifty-second of the time—but cannot predict any single result. Only the concrete act of drawing a card converts potential choice into the hard fact of the ace of spades in your hand. Likewise, in what is called the "collapsing" of the wave function, a physical measurement selects a value inherent in the wave function, to give an actual result. The analogy is inexact because drawings [of cards] are random, whereas experiments measure the effects of physical causes. Nevertheless, the best meaning we have been able to give the wave function is that it represents statistical probabilities. This interpretation simply restates the wave-particle duality; it implies that the act of measurement somehow focuses a wave, which carries all physical possibilities, into a particle with definite properties.[33]

The Schrödinger wave equation[34] evokes the image of an electron as a wave spread out around the nucleus of the atom and not a point particle moving about in space. On the other hand, when the position of an electron is measured, the electron appears at a single point as a particle with mass, negative charge, and spin. It is interesting to note, however, that some physicists would take exception to Professor Perkowitz's statement that the "wave function is a map of potentiality, not a picture of reality." As Professor Stenger notes, "the mathematics [of the Schrödinger wave equation] speaks for itself."[35] By this I think he means that the

---

33. Perkowitz (1996), pp. 85–86.

34. See the previous discussion in this chapter concerning the Schrödinger wave equation.

35. See the quotation from Professor Stenger at the beginning of this chapter.

wave equation is a representation of reality. Professor Penrose also observes that the wave equation is completely deterministic.[36]

The paradox is this: science is based on making measurements to be able to test a theory but the electron wave cannot be directly measured. As stated in the textbook *Fundamentals of Physics:*

> *Only those quantities that can be measured have any real meaning in physics.*[37]

The existence of an electron in the form of a wave has never been measured directly. Neither, however, is it possible to measure the "orbit" of an electron as a particle "orbiting" the atomic nucleus. Remember that, under the rules of classical mechanics, a charged particle "orbiting" an atomic nucleus would radiate light. However, an electron that exists in an energy state around an atomic nucleus does not radiate light. The concept of an electron as a wave was developed specifically to overcome the fact that light emissions are not measured when an electron is in such a state.

Some physicists have used the "collapsing of the wave function" as a possible explanation of human consciousness. They have theorized that human consciousness affects the actions of the atomic particles that make up the human brain. They hypothesize that, based on the wave equation, there are an infinite number of states that an atomic particle can have at any point in time and that human consciousness freezes one of the states and, in effect, "collapses the wave function." Thus, human consciousness is understood as the simultaneous selection of one specific state for trillions of subatomic particles that make up the human brain.

The concept of "collapsing the wave function" is similar to the question as to whether or not a tree falling in the forest makes a sound if no one is around to hear it. The analogous question is

---

36. See the quotation of Professor Penrose in the chapter "Math."
37. Halliday (1974), pg. 789.

"Are there wave functions being collapsed by particles interacting with other particles even when there is no one around to perform measurements?" "What does an electron in an atom look like when no one is looking at it?" Also, "what is special about the measurement apparatus that makes wave functions collapse?" These types of questions have been debated for decades, and we will not resolve them here. For purposes of this analysis, let us assume that wave functions are collapsed, at least by conscious beings.

THE COLLAPSE OF THE WAVE FUNCTION AND FREE WILL

How, then, is the collapsing of the wave function used to try to explain how the brain is able to make free will operate in a human being? The explanation starts with the presumption that a human being is predisposed to a certain course of action and to feelings based on the composite influence of everything that has happened to that person up to any point in time. Then, in each moment in time, the brain selects a specific location and energy level for each subatomic particle from the infinite possibilities for each subatomic particle made possible by the wave function. Each collapse of the wave function rules out all the other possibilities. Together, all the combined actions of the subatomic particles that make up the brain cause the nerves and the synapses of the brain to fire in a certain way, and the brain causes the body to act and think in a certain way. The process is summarized as follows:

> *The wave equation describes an infinite number of possibilities for the position and velocity of the electrons and other atomic particles. The human brain, as it is "thinking" effectively selects one of the wave equation possibilities by "collapsing the wave function" and thus establishes the conditions that will affect how the particles interact with one another. This, in turn, affects the brain neurons and synapses (which are made up of these atomic particles) and thus results in the signals that direct the rest of the body.*

The problem with the above theory is that it does not explain how a brain, which is made out of matter, can have the free will ability to select how the wave functions are collapsed and which neuronal synapses are fired. The attempt to explain how free will can be found in the collapse of the wave function is probably best described by Danah Zohar in the 1990 book *The Quantum Self: Human Nature and Consciousness Defined by the New Physics.* This is similar to the explanations given by David Hodgson in his 1991 book *Mind Matters: Consciousness and Choices in a Quantum World* and by Henry Stapp in his 1993 book *Mind, Matter, and Quantum Mechanics.* Articles on this topic by Mr. Hodgson and Mr. Stapp can also be found in the 1999 book *The Volitional Brain.*[38] Note however, that Mr. Hodgson recognized that:

> *The very fact that decisions [by humans] are made between alternatives suggest an input from something nonphysical.*[39]

Mr. Hodgson went on to hypothesize that the decision-making source of a human is a universal, supernatural soul. This concept is similar to that of Erwin Schrödinger (described later in this chapter) who likewise believed there is one universal soul, rather than individual souls for each human.

As described below, I do not think the attempt to explain the *source* of free will as the collapse of wave functions is successful.

Professor Zohar explains free choice as follows:

> *Each act of concentration is an act of thought realization. Each of us has had the experience that the process of concentration collapses the wave function of a superimposed array of possible thoughts, though few of us might have expressed it this way before being introduced to a quantum vocabulary. By focusing on any one thought, we make that one into a classical reality while the others disappear like so many shadows in the night.*

---

38. Libet (1999).

39. Hodgson (1991), pg. 455.

> *Thus each act of concentration expresses a mini form of choice, a mini form of freedom. Nothing determines upon which one of an array of "possible thoughts" I will focus because the "I" that focuses is itself an indeterminate quantum wave function, but through the act of focusing a choice is made.*[40]

This last sentence reveals that there really is no free choice described by this concept because it is based on the premise that "nothing determines" which thought a human focuses on. In other words, the probabilities inherent in a quantum wave function (which under this concept makes up the "I" in a person) determine what quantum wave forms are chosen, which, in turn, determines what thoughts the person ultimately has or what action is ultimately performed. That is no more a true free choice than to base the choices in life on the flip of a coin or a roll of dice. In Professor Zohar's terminology, "freedom" only means that each particle has an infinite number of potential locations or waveforms based on the wave function. But neither a particle nor a waveform is free to do what it wants.

Professor Zohar then went on to give an example in which her body felt discomfort. She described how she chooses to relieve the tension:

> *... my physical discomfort will spur me to concentrate, and when I do so, I will—by that very act of concentration—choose one possible source of relief for my tension and act on that choice. A choice is, in these terms, nothing but an act of concentration that collapses the wave function of possible thought.*
>
> *But no one can say that **that** specific choice was determined by my discomfort. Any one of my choices would have relieved it. The discomfort only necessitated **some** choice. The choice itself was free.*[41] *[Emphasis in the original.]*

---

40. Zohar (1990), pg. 178.

41. Zohar (1990), pp. 178–179.

It is not clear if Professor Zohar is saying that the act of concentration is a free will act or is a reaction to the discomfort. In the previous quote, she states that the concentration or focus is "indeterminate" because the "I" that is making the choice is subject to the probabilities of a quantum wave function. As stated above, this is not a real "choice." Up to this point, this is the only explanation given for the source of the act of concentration.

Professor Zohar then goes on to explain the mechanism by which the choice is made. Note that this is not a description of the source of the choice but only a description of how it is accomplished or implemented by the brain. She describes the main mechanism that allows the choice to be processed by the brain as being a Bose-Einstein condensate. The Bose-Einstein condensate in the brain is "a coherent ordering of some bosons (photons or virtual photons) present in neural tissue or neuron cell walls."[42] This means there are photons (particles of light) within the brain that interact as a unified group. A laser is the most well-known example of a Bose-Einstein condensate. In a laser, the particles of light do not act independently (what physicists call incoherence) but rather act interdependently and in unison (what physicists call coherence). Professor Zohar explains that "this quantum coherence makes possible the orchestrated firing of some or all of the 10^{11} [that is, 100 billion] neurons in the human brain and the integration of information to which their firing gives rise—thus giving us the unity of consciousness and, ultimately, the sense of self and world."[43]

What Professor Zohar is describing is a system that uses coherent light (the Bose-Einstein condensate) to provide instantaneous communication among all parts of the brain which allows the "orchestrated firing" of brain neurons and simultaneous integration of information across the brain. This allows us to feel

---

42. Zohar (1990), pg. 221.

43. Zohar (1990), pg. 221.

conscious and gives us our sense of self. As described below, the mechanism includes also a "self-reflective capacity of thought" which in engineering terms is best described as a feedback loop. Note that a feedback loop does not provide the source of free will. A feedback loop is a mechanism that measures the output to adjust the internal operations according to a defined program or algorithm.

Professor Zohar explains that a Bose-Einstein condensate can have two states: a low-energy state that acts like a quantum system which operates based on probabilities and a high-energy state that acts like a deterministic classical system with only one condition:

> *This self-reflective capacity of thought to observe itself and thus, through concentration, to collapse its own wave function rests on the physics of at least some Bose-Einstein condensates (including those that are the physical basis of our conscious-ness), on the different physical properties displayed by such quantum systems when they are in a low-energy state or a high-energy state.*
>
> *In a low-energy state, Bose-Einstein condensates display the familiar quantum superposition effects of multiple possi-bilities, experienced by us as the blurry images of our dream life, the Gothic twilight of the imagination. In a high-energy state, these condensates behave almost classically, losing their quantum superposition effects.* [44]

Professor Zohar next explains that this switch of a Bose-Ein-stein condensate in our brain from a low-energy state to a high-energy state is accomplished by the act of concentration.

> *In our conscious system, the act of concentration is the process by which energy is pumped into the brain. We are all aware that when our energy reserves are low, we find concen-*

---

44. Zohar (1990), pg. 179.

tration difficult. But when we do have the energy for concentration, channeling this energy into the brain has the effect of switching the brain's Bose-Einstein condensate from a low-energy quantum state to a high-energy, near-classical state and thus of switching our thought processes from the blurry image of possible thought to the more structured, classical detail of concentrated thought.

On a quantum view of consciousness, then, we have both a basic definition of choice and a basic understanding of the physics that makes choice possible. Any choice, itself, is simply the collapse of the quantum wave function of possible thought into one definite thought, and the physics by which this happens is the switchover of the brain's Bose-Einstein condensate from a many-possibilitied quantum state to a more definite near classical state. All such choices are necessarily free because of the brain's essential quantum indeterminacy—an indeterminacy that exists both in its quantum system and in the firing responses of individual neurons to stimulation.

But this bare-bones model of quantum choice still leaves unanswered all the most interesting questions. How and why, for instance, do I actually make the choices that I do, and if I am free to make any choice, why do I so often make what are clearly bad choices, bad for myself or for others?[45]

Note again that Professor Zohar defines "choice" as the collapse of the quantum wave function. Again, she equates "free" with "indeterminacy," which is not true freedom in the sense that the brain can choose what it wants to do. The collapse of the wave function is not the source of the choice but rather the mechanism used by the brain to implement a choice. This is made clear in the last paragraph in which Professor Zohar explains that the true source of the choice remains unclear. "How and why" she asks "do I actually make the choices that I do?"

---

45. Zohar (1990), pg. 180.

Professor Zohar goes on to explain that it is not the human capacity to reason that determines our choices. Rather, we make a choice as to what we want to do and then explain our choices with reasons. For example, if we choose to give up smoking we explain that we decided to do it because smoking is not healthy. If, on the other hand, we were to choose not to give up smoking, we would likewise have a reason for that decision as well. Possibly we would say that we did not give up smoking because it relieves tension during stressful periods. On this point, I agree completely with Professor Zohar. If our decisions were determined solely by reason and logic, then humans would be no more than complex computers without true free will.

It is this very ability to step back, as it were, to consider two possible choices (and they both might or might not be "reasonable" choices), and then to pick one over the other (based on whatever reason we might want to choose to use) that is the essence of free will.

Professor Zohar quotes from the existential philosophers that, as humans, we have free will and we must make free choices even if we do not have complete knowledge about that which we are making a decision. Dr. Zohar gives further insight into what the source of real choice might be. If a choice is expressed in an act of concentration that collapses a wave function, how is it determined how the wave function will collapse? Professor Zohar further explains:

> *The collapse of a quantum wave function is not random, not wholly without a "sense of direction" . . . Any collapse is a matter of probability, and some outcomes of a collapse are more probable that others. For human quantum systems such as ourselves, the extent to which we can weigh those probabilities is the extent to which we can exercise some control over our freedom.*
>
> *In quantum processes, a probability that something will happen is associated with the amount of energy required to*

make it happen. If an electron can move to one energy shell within the atom with very little expenditure of energy, and to another at very great expenditure of energy, the probability is very high that it will make the low-energy transition. It is free to make any transition, nothing is determined, but it is most likely to take the easy option. So it is with us, though because we are very much more complex characters than electrons, the factors that influence the energy requirements of our various choices are also more complex.[46]

Note that Professor Zohar says that the collapse of the quantum wave function is normally based on probabilities as one would expect for something that follows the laws of quantum physics. But how do humans "exercise some control?" She says it is by "weighing" the probabilities. Apparently this means that humans can somehow choose how a wave function will collapse even if it means going against what might be expected based on the laws of physics without some "outside" intervention. For a dramatic example of "weighing the probabilities" and choosing a low-probability action that requires a lot of effort and focused attention, see the discussion on free will solutions to the problem of obsessive-compulsive behavior in the chapter "The Soul-Brain Interface."

At this point, let us summarize the mechanism of free will proposed by Professor Zohar as described above:

- Choice is an act of concentration, which pumps energy to the Bose-Einstein condensate in the brain, causing it to go into a higher energy level that collapses the wave function.
- The collapse of the wave function is based on probabilities.
- Humans can weigh those probabilities to determine which choice to make.

---

46. Zohar (1990), pg. 183.

So in which of the above three steps can the real source of human free will choice be found? Can it be found in the act of concentration? Professor Zohar does not provide an explanation of how humans can choose what they will concentrate on. In the example she cited, her bodily discomfort made her concentrate on finding a solution. But can she exercise free will and choose to ignore her bodily discomfort? If so, she does not provide an explanation of how this is accomplished.

The source of human free will cannot be in the collapse of the wave function because the collapse of the wave function is based on probabilities. Actions based on probabilities provide no more free will than the roll of dice. Also, human free will cannot be explained by the probabilities of the wave function because Professor Zohar claims that humans can "weigh" those probabilities. As Professor Zohar says:

> *We are always free to choose against the weight of probability, to make more energy-demanding choices, and this freedom makes us responsible.*[47]

So it appears that the ability to "weight the probabilities" and "choose against the weight of probability" is the mechanism that provides free will choice. By causing energy to move the Bose-Einstein condensate into a higher energy state, humans are able to go against what would be the normal probabilities inherent in the wave function. But what then is the source of this ability to "weigh" the probabilities of the wave function and then to "choose against the weight of probability?" Here is where the quantum physics model of human free will reaches its limit. Since atoms and molecules in the brain interact according to quantum probabilities, the ability to "choose against the weight of probability" must come from outside the brain. Quantum physics only provides the mechanism for implementing a choice or a decision.

---

47. Zohar (1990), pg. 185.

Quantum physics is not the source of the choice or the source of the decision.

Professor Zohar does not provide an explanation as to how humans are able to initiate a choice to act against the weight of probability. Professor Zohar very correctly describes that "the stories of people who have risen above their backgrounds or circumstances to do surprising or great things inspire us just because they remind us that we, too, *could* act against the odds, that the responsibility for this lies with no one but ourselves."[48] [Emphasis in the original.] She also explains that "[w]e have a moral imperative to use our freedom, to live at the fearsome edge of our consciousness when called upon to do so, because it is our very nature as conscious beings to be free, and in quantum terms the natural and the moral go hand in hand."[49]

These are very excellent principles by which to live, but they do not explain the source of our freedom and ability to make free will choices. As Professor Zohar asks: "How and why do I actually make the choices I do?" What entity is "weighing" the probabilities and what entity is making the decision to "go against the weight of probability?" It is only by something that is outside the natural world, that is not subject to the probabilistic laws of quantum physics or any other laws of physics that we are able to "weigh the probabilities" and "go against the weight of probability" to make the choices we want to make. It is only by something that is "above" nature, by something that is supernatural that humans are able to exercise free will and make real choices.

As described later under the heading "Proto-consciousness," Professor Zohar has another concept which is purported to describe human consciousness. It, however, also fails to provide an explanation of human free will.

---

48. Zohar (1990), pg. 185.
49. Zohar (1990), pg. 186.

Probabilities and Collapsing the Wave Function Cannot Explain Free Will

Let us examine whether the aforementioned explanations provide a logical basis for understanding the operation of human free will. In the simplest terms, the above explanations do not successfully explain free will because they rely on circular reasoning. If the human brain is made up of atomic particles and wave functions that are not affected by an outside force or decision-making agent, then any "thinking" that is going on is simply the interaction of those particles and wave functions. Based on the theory of quantum mechanics, atomic particles and wave functions will certainly interact with one another. But they interact based on the probabilistic laws of quantum physics. As described in this chapter, quantum mechanics can be used to arrive at very predictable results. The results are so predictable that, after decades of experimentation, no experiment has been found that contradicts the theory. Thus if "thinking" and "concentrating" are just the interaction of atomic particles and wave functions, then the "thinking" and "concentrating" will be governed by the probabilistic laws of quantum physics. Such "thinking" and "concentrating" will not be free.

A probabilistic model can be described by an example. Suppose there is a large auditorium with 1 million pennies held in buckets near the ceiling. If all the pennies were dropped to the floor, the number of heads and the number of tails would be within a very small percentage of 500,000 each (50 percent heads and 50 percent tails). If all of the pennies were numbered before they were dropped, it would be impossible to predict for any given penny whether it would end up "heads" or "tails." In fact, if you were to try to predict how any specific penny would land, there would be a 50 percent chance that you would be wrong. Conversely, you would be able to predict with a high degree of accuracy the percentage of all pennies that would land as "heads" and the percentage that would land as "tails."

This example describes how a probabilistic model works. Note that even though the model is not deterministic and we cannot predict the outcome for a specific coin, the outcome overall is very predictable. There is a very small amount of unpredictability in each drop as to whether heads or tails will be greater than 50 percent. Also, if we continue to make drop after drop, the total percentage of heads and tails for all drops will come closer and closer to 50 percent unless there is some outside force affecting the coins. The coins cannot decide, either individually or as a group, to make themselves turn up heads or tails. The coins cannot decide, for example, to make themselves turn up 60 percent heads and 40 percent tails.

But, you may say, through the senses the human brain receives stimulus from outside forces that can change the probabilities of interaction. Here it is necessary to make a distinction between an outside force that simply brings information into the brain and an outside force that is an agent for making decisions. The senses bring information into the brain in the form of electrochemical impulses. The brain must interact with these impulses following the laws of physics. If there is no outside force that is an agent for making decisions, then humans are nothing more than very complex biological computers that receive "input" via the senses, process the information in a complex biological central processor (the brain) according to the brain's hardware (genetics or nature) and software (upbringing or nurture). The brain then directs our actions based on the hardware and software.

To be able to explain free will with the uncertainty principle, the electrons and other subatomic particles would somehow have to have minds of their own. For electrons to have minds of their own, they would have to move in ways that do not follow the laws of physics. But as described above, this has never been observed to happen. Thus any theory which claims that inanimate atomic particles are the source of human free will contradicts decades of scientific evidence.

The probabilistic model can be applied to the human brain. There are trillions of atomic particles in the human brain. Based on scientific principles, all the particles move and interact with the other particles according to the laws of quantum physics. None of the particles can "decide" to move or interact in ways that would violate the laws of physics. According to the quantum theory, it is impossible to predict the exact location and velocity of individual particles, but the composite interaction of the particles is welldefined. The quantum theory does not provide flexibility of a type that is able to explain human free will.

PHYSICAL MANIFESTATIONS OF CONSCIOUSNESS

Many authors have recognized the apparent connection between physical states of the brain and consciousness. For some, this indicates that consciousness can be or will ultimately be explained as a physical phenomenon. Others recognize the inherent difficulty in explaining how free will and consciousness could be merely the interaction of inanimate atoms and molecules. Professor Weinberg provides an excellent summary of the two viewpoints:

> *Of all the areas of experience that we try to link to the principles of physics by arrows of explanation, it is consciousness that presents us with the greatest difficulty. . . . The physicist Brian Pippard, who held Maxwell's old chair as Cavendish Professor at the University of Cambridge, has put it thus: "What is surely impossible is that a theoretical physicist, given unlimited computing power, should deduce from the laws of physics that a certain complex structure is aware of its own existence."*
>
> *I have to confess that I find this issue terribly difficult, and I have no special expertise on such matters. But I think I disagree with Pippard and the many others who take the same position. It is clear that there is what a literary critic might call an objective correlative to consciousness; there are physical and chemical changes in my brain and body that I observe*

*to be correlated (either as cause or effect) with changes in my conscious thoughts. I tend to smile when pleased; my brain shows different electrical activity when I am awake or asleep; powerful emotions are triggered by hormones in my blood; and I sometimes speak my thoughts. These are not consciousness itself; I can never express in terms of smiles or brain waves or hormones or words what it **feels** like to be happy or sad. But setting consciousness to one side for a moment, it seems reasonable to suppose that these objective correlatives to consciousness can be studied by the methods of science and will eventually be explained in terms of the physics and chemistry of the brain and body.[50] [Emphasis in the original]*

I, of course, believe Professor Pippard is correct. Professor Weinberg's main reason for believing that consciousness is solely a physical phenomenon is that there is a strong correlation between brain chemistry and how we feel. I do not deny this correlation, but I would amplify Professor Weinberg's own parenthetical comment: the physical manifestation could either be the cause of how we feel or the physical manifestation could be affected by what we are thinking or feeling. It also seems clear that this is a very complex phenomenon such that at times our feelings might begin with a physical cause but then could be affected by how we choose to interpret those feelings. Conversely, our feelings might begin with an independent thought that causes a hormonal reaction which influences how we feel. Thus even though there is undoubtedly a physical component to our feelings, there could also be a supernatural source that is either the initiator of feelings or at least the interpreter of physical feelings.

Since all feelings will have some physical manifestation, I recognized the inherent difficulty in providing evidence that there might be a supernatural component of feelings as well. Thus I have chosen instead to focus on the concept of free will which I do not think can be explained by physical matter that must follow the laws of physics. Although all consciousness seems to embody

---

50. Weinberg (1992), pg. 44.

some kind of feeling, more fundamentally, human consciousness is characterized by free will.

Professor Weinberg continues on from the previous quotation to describe what he means to "explain" consciousness:

> *(By "explained" I do not necessarily mean that we will be able to predict everything or even very much, but that we will understand why smiles and brain waves and hormones work the way they do, in the same sense that, although we cannot predict next month's weather, still we understand why the weather works the way it does.) . . . It is not unreasonable to hope that when the objective correlatives to consciousness have been explained, somewhere in our explanations we shall be able to recognize something, some physical system for processing information, that corresponds to our experience of consciousness itself, to what Gilbert Ryle has called "the ghost in the machine."[51] That may not be an explanation of consciousness, but it will be pretty close.[52]*

In this statement, Professor Weinberg seems to imply that consciousness (and presumably free will) is similar to the weather, in the sense that both are caused by the laws of physics. However, both are so complex that it would not be possible to write an equation or a computer program that would allow us to predict what either will do in the future.[53] I do not agree with this characterization of either consciousness or free will. It is important to realize that free will is not just based on the inability to predict what something will do in the future due to the fact that the causative factors are too complex. Rather, it is not possible to predict what a human with free will will do because that human is able to

---

51. See the chapter "The Soul-Brain Interface" for an additional comment about Professor Ryle. He considered the concept of a "ghost in a machine" to be the failed concept of a supernatural soul.

52. Weinberg (1992), pg. 45.

53. Such a description of free will is comparable to Professor Stephen Hawking's explanation of free will as described in the chapter "Free Will or Not."

make choices and those choices are not controlled by the laws of physics. It is the ability to make free choices that defines free will. The inability to model a complex interaction that is governed completely by the laws of physics is not a description of free will.

Schrödinger on "What Is Life"

In 1926, Professor Erwin Schrödinger developed the wave equation which forms the basis for quantum physics and modern chemistry. Eighteen years later, in 1944, Professor Schrödinger wrote a book entitled *What Is Life*, which addressed the question of reconciling the laws of physics with the biological discoveries concerning chromosomes. Of interest to our discussion on human souls, in an Epilogue entitled "On Determinism and Free Will," Professor Schrödinger concluded that matter alone cannot explain free will:

> . . . *let us see whether we cannot draw the correct, non-contradictory conclusion from the following two premises:*
>
> *My body functions as a pure mechanism according to the Laws of Nature.*
>
> *Yet I know, by incontrovertible direct experience, that I am directing its motions, of which I foresee the effects, that may be fateful and all-important, in which case I feel and take full responsibility for them.*
>
> *The only possible inference from these two facts is, I think, that I—I in the widest meaning of the word, that is to say, every conscious mind that has ever said or felt "I"—am the person, if any, who controls the "motion of the atoms" according to the Laws of Nature. . . . it is daring to give to this conclusion the simple wording that it requires. In Christian terminology to say: "Hence I am God Almighty" sounds both blasphemous and lunatic. But please disregard these connotations for the moment and consider whether the above inference is not the closest a biologist can get to proving God and immortality at one stroke. . . . From the early great Upanishads, the recogni-*

> *tion ATHMAN = BRAHMAN (the personal self equals the omnipresent, all-comprehending eternal self) was in Indian thought considered, far from being blasphemous, to represent the quintessence of deepest insight into the happenings of the world. . . . The mystics of many centuries, independently, yet in perfect harmony with each other (somewhat like the particles in an ideal gas) have described, each of them, the unique experience of his or her life in terms that can be condensed in the phrase: DEUS FACTUS SUM (I have become God). . . . To Western ideology the thought has remained a stranger . . . .[54]*
> *[Upper case in the original.]*

I agree with Professor Schrödinger that it is incompatible to say both that human beings are only made up of a material body and that human beings have free will. I do not, however, agree with him that the only logical conclusion is that each of us is "God Almighty." He fails to consider that each of us might have a unique supernatural soul that controls "the motion of atoms" in only a limited space, that space being inside the skull of one human body. I can understand why he thinks that the ability to affect the motion of atoms is God-like. Moving matter with a supernatural force is one of the images we normally associate with God. God is often characterized as the ultimate uncaused Cause of the universe. Humans, however, can have a God-like power that enables them to be the uncaused cause of their own free choices. As explained in the chapter "Introduction and Basic Premise," the interactions of all natural phenomena are subject to prior causes that affect how they interact. An independent, uncaused cause, such as a human free choice, must have a supernatural source. Human power, however, is infinitesimally puny compared to the power of an almighty, spiritual being who can create matter and spiritual beings from nothing and who can integrate spiritual souls with material bodies.

---

54. Schrödinger (1944), pps. 86–87.

Apparently, Mr. Schrödinger discounts the possibility of multiple human souls because he goes on to describe how some people in history have asked whether or not animals or women have souls:

> *How does the idea of plurality (so emphatically opposed by the Upanishad writers) arise at all? Consciousness finds itself intimately connected with, and dependent on, the physical state of a limited region of matter, the body. (Consider the changes of mind during the development of the body, as puberty, aging, dotage, etc., or consider the effects of fever, intoxication, narcosis, lesion of the brain, and so on.) Now, there is a great plurality of similar bodies. Hence the pluralization of consciousnesses or minds seems a very suggestive hypothesis. Probably all simple, ingenuous people, as well as the great majority of Western philosophers, have accepted it.*

> *It leads almost immediately to the invention of souls, as many as there are bodies, and to the question whether they are mortal as the body is or whether they are immortal and capable of existing by themselves. The former alternative is distasteful, while the latter frankly forgets, ignores, or disowns the facts upon which the plurality hypothesis rests. Much sillier questions have been asked: Do animals also have souls? It has even been questioned whether women, or only men, have souls.*

> *Such consequences, even if only tentative, must make us suspicious of the plurality hypothesis, which is common to all official Western creeds. Are we not inclining to much greater nonsense, if in discarding their gross superstitions we retain their naive idea of plurality of souls, but "remedy" it by declaring the souls to be perishable, to be annihilated with the respective bodies?*

> *The only possible alternative is simply to keep to the immediate experience that consciousness is a singular of which the plural is unknown; that there is only one thing and that what seems to be a plurality is merely a series of different aspects of*

this one thing, produced by a deception (the Indian MAJA); the same illusion is produced in a gallery of mirrors, and in the same way Gaurisankar and Mt. Everest turned out to be the same peak from different valleys.[55] *[Upper case in the original.]*

It appears Professor Schrödinger discounts the possibility of multiple souls because some people have asked what he considers to be silly questions such as whether or not souls are immortal or whether or not animals have souls. There are undoubtedly many people who would not consider those questions silly. Nevertheless, if it is appropriate to discount any approach to understanding life because at one time or another someone has asked silly questions about it, we would probably have to discount all human knowledge and experience, including all of science.

I have yet to understand the logic of there being only one Consciousness of which we are all a part. If each of us is a part of the same soul, why are there differences of opinion, why do we hurt each other, and how can we do something that the Consciousness does not want us to do? For example, how can people do things that are evil if we are all controlled by the same Consciousness? What is the source of free will and willful choice if we are all part of the same soul? In my opinion, if indeed each human has free will, each human must therefore have his or her own soul which is the source of that free will. I consider it only semantics to say that we are all part of the same soul.

Proto-consciousness

According to Danah Zohar, who rejects the concept of supernatural souls, there is an explanation for free will which is based on the natural world. In the 2000 book *SQ Connecting with Our Spiritual Intelligence*, written with her husband Dr. Ian Marshall, Professor Zohar describes the concept of proto-consciousness. By this, she means that all matter has the quality of consciousness

---

55. Schrödinger (1944), pg. 89.

but that it is only inside the brains of animals that it is organized in such a way that it acts consciously. Professor Zohar describes the hypothesis of proto-consciousness:

> *In this view, proto-consciousness is a natural part of the fundamental physical law of the universe and has been present since the beginning of time. Everything that exists—fundamental particles like mesons and quarks, atoms, stones, stars, tree-trunks and so on—possesses proto-consciousness.*[56]

Professor Zohar accepts the concept of proto-consciousness:

> *I back this proto-consciousness view. It makes no sense to me that consciousness should appear just arbitrarily out of nowhere. Equally, it seems too strong to suggest that things like atoms and stones are conscious in the way that we are conscious. The notion that "brute matter" possesses a weak form of proto-consciousness that becomes full-blown consciousness only in certain structures like brains seems to fit sensibly in between. It has a ring of credibility. Yet even a theory like this has a missing link. We still need to propose some sort of bridging phenomenon in the brain from proto-conscious brute matter to single neurones and then on to fully conscious coherent neural oscillations. To do this, I feel strongly that we must look at quantum phenomena in the brain. These may supply the necessary bridge and show why brains have what it takes to generate full-blown consciousness.*[57]

First of all, I agree with Professor Zohar that "it makes no sense . . . that consciousness should appear just arbitrarily out of nowhere." However, to me, it makes sense that consciousness and free will in humans has a supernatural rather than a natural origin. As described in this chapter, matter, which must follow physical laws cannot have the freedom to be the source of free will decisions and choices.

---

56. Zohar (2000), pg. 81.
57. Zohar (2000), pg. 83.

Professor Zohar recognizes that humans are conscious and can make free will choices. However, she also recognizes that atomic particles cannot make free will choices as shown by countless scientific experiments. Nevertheless, she assumes that atomic particles have something called "proto-consciousness." As described below, this proto-consciousness is just another name for the indeterminate nature of the wave function which describes the actions of atomic particles. As described in great detail in this chapter, the indeterminate nature of quantum physics does not provide a source for free will.

Professor Zohar claims that it is the Bose-Einstein condensate that provides the bridge required between "proto-conscious brute matter" and "full-blown consciousness:"

> *Evidence for coherent states (Bose-Einstein condensates) in biological tissue is now abundant, and the interpretation of its meaning is at the cutting edge of exciting breakthroughs in our understanding of what distinguishes life from nonlife.* **I think that the same Bose-Einstein condensation among neuron constituents is what distinguishes the conscious from the nonconscious. I think it is the physical basis of consciousness.**[58] *[Emphasis in the original.]*

As described under the heading "The Collapse of the Wave Function," the Bose-Einstein condensate may well provide the physical mechanism for the implementation of free will decisions and consciousness, but it is not the source of free will. Note that Professor Zohar rejects the notion that atomic particles can somehow make free will decisions. In discussing a photon's apparent ability to "choose" between being a wave or a particle in certain experiments, Professor Zohar admits that:

> *I very much doubt that photons actually make choices.*[59]

---

58. Zohar (1990), pg. 85.
59. Zohar (1990), pg. 197.

This, of course, is completely consistent with decades of experimental results. It is perfectly predictable under which experiments the wave characteristics of photons are seen and under which experiments the particle characteristics of photons are seen. Photons do not have the ability to choose when they will exhibit their wave characteristics and when they will exhibit their particle characteristics. The same can be said of experiments with subatomic particles, such as electrons, which have the characteristics of both waves and particles.[60] As described at the beginning of this chapter, the quantum theory has been the most successful theory in science for predicting experimental results. Even one of the quotes attesting to the success of quantum theory is by Professor Zohar.

Likewise, Professor Zohar disparages the notion that electrons can make free will choices (panpsychism is similar to proto-consciousness as described below):

> *It is also true, and certainly relevant when trying to make any panpsychist resolution of the mind/body problem sound convincing, that panpsychism in almost any form so far conceived is vaguely embarrassing. It makes people shift in their seats. Even when disclaimers like "very primitive," "elementary," and "proto-" are used to discuss the consciousness of elementary particles, one can't help conjuring up images of electrons falling in love or fretting about whether they might not perform well in their next two-slit experiment.[61]*

Panpsychism is the belief that matter and spirit are inseparable and that everything exhibits the qualities of both, in one degree or another. Scientific experiments over the last several hundred years have, of course, provided evidence that completely

---

60. For those interested, most physics text books describe the experiments under which atomic particles exhibit characteristics of waves and particles. These include the two-slit experiment referred to by Professor Zohar.

61. Zohar (1990), pg. 97.

contradicts this belief. All scientific evidence to date indicates that matter and energy interact according to laws.

Panpsychism, however, was developed long before the scientific revolution of the last several hundred years:

> *It was clear in our discussion of the possibility that electrons might be conscious that panpsychism of one sort or another has appealed to philosophers and scientists since the beginning of recorded thought. It has colored the thinking of people as widely separate as Parmenides and Heraclitus, Spinoza, Whitehead, and Bohm. Its attraction, like that of materialism, lies in the wish to find one unifying substance that undercuts all divisions of the world into the mental and the material. Unlike materialism or idealism, it tries to do so without denying the reality of either.*[62]

If Professor Zohar does not believe atomic particles can make free will choices, what is the nature of the proto-consciousness of atomic particles? Professor Zohar, as well as other modern physicists, apparently have resurrected the concepts of panpsychism and restyled them in quantum mechanical language. The indeterminate nature of the wave function in atomic particles is equated with the presumed proto-conscious quality of atomic particles. But as discussed previously in this chapter, this indeterminacy does not provide a basis for free will. I do not agree that there is a "ring of credibility" that "proto-conscious" atomic particles, which do not individually have the capacity for free will decisions, can somehow be organized to make free will decisions. Professor Zohar has not provided any explanation or conceptual model of how particles that cannot individually make free will decisions can be organized in such a way that they can make free will decisions as a unified organism. There is no scientific evidence that any matter has proto-consciousness. On the contrary, there are decades of scientific research that indicate matter always follows scientific laws.

---

62. Zohar (1990), pg. 96.

In her 1990 book *The Quantum Self,* excerpted previously, Professor Zohar appears to discredit panpsychism as "vaguely embarrassing" even when disclaimers like "proto-" are used to "discuss the consciousness of elementary particles." She describes how it implies the nonsensical image of electrons "falling in love or fretting about whether they might not perform well in their next two-slit experiment." This is the very same panpsychism to which she refers as proto-consciousness and which she embraces as having a "ring of credibility" in her 2000 book *SQ.* I could find no explanation for her apparent change of mind.

NONREDUCTIVE PHYSICALISM

The concept of nonreductive physicalism is also used by some philosophers to attempt to explain free will as a natural phenomenon. Nonreductive physicalism is described in a series of articles by several authors compiled in the 1998 book *Whatever Happened to the Soul?* Nonreductive physicalism is similar to the concept of proto-consciousness as described above. To provide a conceptual basis for free will, nonreductive physicalism relies on the assumption that the matter that makes up the brain can somehow have a mind of its own and not follow the laws of physics. This is a very nonscientific assumption. Decades of scientific experiments indicate that all natural phenomena follow the laws of physics. The significant advancements made possible by science since the 1600s have been due to the ability of scientists to explain more and more of the natural world using scientific laws. The assumption that the matter inside the brain is nonreductive and does not follow the laws of physics would bring us full circle back to the prescientific days when people did not realize that the natural world does follow the laws of physics. In current times, when there is a situation in which it is theorized that matter does not follow the laws of physics, it is referred to as a miracle.

Nancey Murphy, one of the contributing authors to *Whatever Happened to the Soul?*, believes that scientific research that has shown a correlation between brain activity and "mental" phenomenon provides strong evidence that the matter in the brain is not reducible to the laws of physics.[63] However, that is circular reasoning. If there are supernatural components of "mind," it is possible they would work through the brain and thereby result in correlated brain activity.[64] Thus brain activity correlated to mental activity does not provide evidence that mental activities (including free will) have a natural rather than a supernatural origin.

Whatever Happened to the Soul? does not contain a conceptual theory as to how matter that is inside the brain could free itself from the laws of physics to have a mind of its own and to allow for free decisions to be made.

Beyond Quantum Mechanics and Relativity— The Final Theory

Despite the great success of Albert Einstein's theory of relativity in explaining the nature of gravity and how it affects the motion of cosmic bodies and the success of quantum mechanics in describing atomic interactions, there remains one huge problem: the theories are not compatible. As described by Professor Weinberg in his 1992 book *Dreams of a Final Theory:*

> *The principles of relativity and quantum mechanics are almost incompatible with each other and can coexist only in a limited class of theories.*[65]

In his 1997 book, *The Life of the Cosmos*, Professor Lee Smolin provides a general explanation as to why the two theories are not compatible:

> *Quantum theory has yet to successfully encompass the phenomena of gravitation, and Einstein's theory of relativity can*

---

63. Brown (1998), pg. 140.

64. See the chapter "The Soul-Brain Interface" for a description of several theories as to how a supernatural entity might be able to interact with a physical brain.

65. Weinberg (1992), pg. 24.

only explain gravitation by ignoring quantum theory and treating matter as if Newton's worldview still held.[66]

In his 1999 book *The Elegant Universe: Superstrings, Hidden Dimensions, and the Quest for the Ultimate Theory*, physics professor Brian Greene describes the conflict between quantum mechanics and general relativity as follows:

The notion of a smooth spatial geometry, the central principle of general relativity, is destroyed by the violent fluctuations of the quantum world on short distance scales.[67] *[Emphasis in the original.]*

The following is a short discussion of the background leading to this conflict between quantum mechanics and general relativity. There are four forces that have been identified as the fundamental forces in the universe. The nature of these forces is described below, in order of increasing strength.

Gravity:	Force that attracts any two masses (in Newtonian terms) or Force that deforms space and time (in Einsteinian terms)
Weak nuclear force:	Force that causes nuclear particles to be expelled from the nucleus of the atom. These expelled particles are known as radioactivity.
Electromagnetic force:	Force associated with electrically charged particles and magnetic fields. Opposite charges and fields attract. Like charges and fields repel. Maxwell's equations describe the interrelationships.

---

66. Smolin (1997), pg. 5.
67. Greene (1999), pg. 129.

Strong nuclear force: Force that binds together the nuclear particles of an atom. Inside the nucleus, this force is much stronger than the electromagnetic force between the positively charged protons in the nucleus that would tend to make them repel each other. Outside the nucleus, the strength of this force is essentially zero.

Albert Einstein spent the last 30 years of his life trying to reconcile the electromagnetic force with the theory of relativity (which describes gravity) in a unified field theory. He was not successful.[68]

There has been success, however, in unifying some of the other fundamental forces. As described earlier, James Maxwell unified the electric and magnetic forces into the electromagnetic force in the 1860s. In the early 1970s, Professors Weinberg, Sheldon Glashow, and Abdus Salam developed a theory that unified the electromagnetic and weak nuclear force into the electroweak force. In 1979, they were awarded the Nobel prize in physics for their work and, in the 1980s, the key predictions of the theory were confirmed in high-energy experiments. According to their theory, the two forces are the same at high energy levels, but at lower energy levels, they are different manifestations of the same force.

More recently, physicists have proposed grand unifying theories (GUTs) that would unify the electroweak and the strong nuclear force (but not gravity). Unfortunately, testing the GUTs is not practical. Professor Hawking estimates (and jokes) that "a machine that was powerful enough to accelerate particles to the grand unification energy would have to be as big as the Solar System—and would be unlikely to be funded in the present

---

68. Weinberg (1992), pg. 18.

economic climate."[69] The goal of physicists is to ultimately develop one consistent theory that unifies all of the above forces (including gravity) in a noncontradictory way. This theory is usually referred to as a "Theory of Everything" or "TOE." Professor Stephen Hawking has characterized the TOE as the "mind of God" because it would describe the entire natural universe.[70] To better understand the complications associated with achieving a TOE, the reader should consult the books by Professors Hawking, Smolin, Weinberg, and Greene.

I would like to comment that although physicists might ultimately be successful in developing a consistent TOE which describes the entire natural universe, the TOE will not describe human souls or the spirit of God. Thus it will not be able, even conceptually, to predict the future decisions and free will choices of humans. The TOE will know only the "mind of God" concerning the natural universe, and it will not know the mind of God about spiritual matters such as how humans should live their lives and what choices humans should make to live morally good lives. The laws of physics do not provide moral guidance. This is one of the points that the philosopher Immanuel Kant made in his 1781 book *A Critique of Pure Reason*.

The model now accepted by the scientific community is known as the standard model. This model incorporates the merging of the electromagnetic force and the weak nuclear force into the electroweak force described above. It also incorporates a theory known as quantum chromodynamics which describes the strong nuclear force in terms of various types of quarks which are bound together (with gluons) to make protons and neutrons. Quarks are smaller subatomic particles that cannot be found in isolation but must always be found bound together. The standard model has met all experimental tests:

---

69. Hawking (1988) 1996 ed., pg. 76.
70. Hawking (1988) 1996 ed., pg. 191.

. . . on the experimental side, nothing has been discovered that could not be explained in terms of the standard model.[71]

As described at the beginning of this chapter, the above statement was true up until 2001 when scientists performing a high-energy experiment discovered a muon that could not be explained by the standard model. The motion of the muon, however, was consistent with string theory.

One of the main candidates for unifying all of the fundamental particles and forces (including gravity) is called string (or superstring) theory. This theory is based on fundamental particles in the shape of strings that are billions of times smaller than neutrons and protons. The strings vibrate in various modes and are joined in various configurations to make up the fundamental forces and particles. One of the string theories envisions a nine-dimensional world in which six of the dimensions are "rolled up" resulting in our familiar three-dimensional world.

There is a fundamental problem with testing any string theory. The experimental energy needed to break atomic particles apart to reveal the underlying strings is not realistically achievable:

> *The really fundamental Planck energy where all these questions could be explored experimentally is about a hundred trillion times higher than the energy that would be available at the Superconducting Super Collider.*[72] *It is at the Planck energy where all the forces of nature are expected to become unified. Also, this is roughly the energy that, according to modern string theories, is needed to excite the first modes of vibration of strings, beyond the lowest modes that we observe as ordinary quarks and photons and the other particles of the standard model.*[73]

---

71. Smolin (1997), pg. 33.

72. The Superconducting Super Collider was a high-energy particle accelerator that was proposed in the 1990s but never received full funding. Some tunnels were built in Texas, but the project was abandoned.

73. Weinberg (1992), pg. 234.

Nevertheless, physicists have come up with some novel experimental approaches they hope will provide some evidence for the string theory. One approach hopes to use a particle accelerator to measure the energy of gravitons (or gravity particles) which are predicted by one of the string theories. Of course, it is not possible to predict how or if ever, a final consistent theory can be achieved that will explain or unify all of the fundamental forces of the universe.

Regardless of which theory is ultimately successful, the conclusions reached in this book would not be affected. None of the theories include particles, waves, or forces that "have minds of their own" and can choose when to follow the laws of physics and when not to follow them. A theory that contained particles, waves, or forces that "have minds of their own" would be decidedly non-scientific. As described above, panpsychism and its modern derivative, "proto-consciousness," are examples of such a nonscientific theory that presumes matter has both material and supernatural characteristics. However, the theories of panpsychism and proto-consciousness are not consistent with the evidence from decades of scientific research.

As explained by Professor Greene, whose book, *The Elegant Universe*, covers the realms of physics ranging from string theory (which deals with the smallest fundamental phenomena yet envisioned) to the theory of general relativity (which covers phenomena of cosmic extent):

> *Physicists—and most everyone else as well—rely crucially upon the stability of the universe: The laws that are true today were true yesterday and will still be true tomorrow (even if we have not been clever enough to have figured them all out). After all, what meaning can we give to the term "law" if it can abruptly change? . . . The simplest assumption that is consistent with all that we know is that the laws [of nature] are fixed. . . . Thankfully, everything we know points toward the laws of physics being the same everywhere. All experiments*

*the world over converge on the same set of underlying physical explanations. Moreover, our ability to explain a vast number of astrophysical observations of far-flung regions of the cosmos using one, fixed set of physical principles leads us to believe that the same laws **do** hold true everywhere.*[74]

WHAT IS REALITY?

Occasionally, after he had studied too long at one time, my roommate in college would scream out "What Is Reality?". After studying about the nature of quantum mechanics and string theory, one might be tempted to scream the same thing. Note that the questions raised below are some of the questions quantum physicists have discussed and continue to discuss without necessarily coming to a "final" answer.

Ever since the wave equation was developed by Erwin Schrödinger in 1926, physicists have debated the nature of reality at the atomic level. As Professor Perkowitz comments, according to the interpretation developed by Max Born in 1926, "[T]he wave function is a map of potentiality, not a picture of reality." Professor Weinberg also explains:

> . . . *electron waves are not waves of anything; their significance is simply that the value of the wave function at any point tells us the probability that the electron is at or near that point.*[75] *[Emphasis in the original.]*

This interpretation implies that an electron is not "really" a wave, but it is a particle that can be found with various levels of probability in areas around the nucleus of an atom.[76] This interpretation does not appear to address the problem the wave equation was designed to solve. If an electron is a particle, how does the electron move around the atomic nucleus without sending

---

74. Greene (1999), pg. 168.

75. Weinberg (1992), pg. 72.

76. Some electron shells are spherical, others have much more complex shapes.

out electromagnetic radiation? Or is the electron a wave until we look at it? Again, Professor Weinberg:

> *As Max Born had emphasized, during the times between measurements, the values of the wave function evolve in a perfectly continuous and deterministic way, dictated by some generalized version of the Schrödinger equation.... Quantum mechanics itself is not vague; although it seems weird at first, it provides a precise framework for calculating energies and transition rates and probabilities.*[77]

If the wave function is not "real," how can it keep evolving between measurements? If the wave function is not "real," why do we talk about it collapsing? Does a wave function really collapse? If so, how often does it collapse? Does it only collapse when we are looking at it? What role does a conscious being play in causing a wave function to collapse? Do photons from sources not controlled by conscious beings result in wave functions collapsing? Why should the measurement process, which is only some atomic particles interacting with other atomic particles (or some atomic wave functions interacting with other atomic wave functions), be anything different than what an electron is "normally" doing? Professor Weinberg describes it well when he says:

> *. . . this difference of treatment between the system being observed and the measuring apparatus is surely a fiction. We believe that quantum mechanics governs everything in the universe, not only individual electrons and atoms and molecules but also the experimental apparatus and the physicists who use it. If the wave function describes the measuring apparatus as well as the system being observed and evolves **deterministically** according to the rules of quantum mechanics even during a measurement, then . . . where do the probabilities come from?*[78] *[Emphasis in the original.]*

---

77. Weinberg (1992), pg. 75.
78. Weinberg (1992), pg. 81.

Note that Professor Weinberg provides a strong case for saying that quantum mechanics represents a deterministic model of the universe, even when measurements are taking place. The astrophysicist Stephen Hawking has a similar view:

> *These quantum theories are deterministic in the sense that they give laws for the evolution of the wave with time. The unpredictable, random element comes in only when we try to interpret the wave in terms of the positions and velocities of particles. But maybe that is our mistake: maybe there are no particle positions and velocities, but only waves. It is just that we try to fit the waves to our preconceived ideas of positions and velocities. The resulting mismatch is the cause of the apparent unpredictability.*[79]

What is interesting about the preceding statements is that they appear to say that the waves are real and that the particles are not, whereas other statements appear to say just the opposite. Also, the waves are deterministic whereas the particle positions and velocities are not. I myself favor the wave explanation over the particle explanation because the wave explanation was originally developed to overcome the inconsistencies of trying to explain electrons as particles that circle around the nucleus of an atom. I like Professor Perkowitz's concept that the measurement process "focuses" the wave of an electron. Note that when an electron is not near an atom, it is a "bundle of the energy of a field" (as described by Professor Weinberg). Thus, I prefer to refer to the measurement of an electron as a rebundling of the electron wave rather than as a wave function collapse. This may be a partial explanation of "where the probabilities" of quantum mechanics come from. We cannot "see" the electron in its wave-form around the atom without detecting it with a photon. But the photon interferes with the conditions that make up the wave equation that describes the electron around the atomic nucleus. This results in the electron being found rebundled as an electron

---

79. Hawking (1988) 1996 ed., pg. 189.

wave bundle at a specific place, rather than as a standing wave around the atomic nucleus. The study of the interaction of photons and electrons is know as quantum electrodynamics.

However, whether matter is comprised of waves or particles, the matter that makes up the human brain is not "free" to act as it wants such that it can explain free will. Whether quantum mechanics is deterministic or probabilistic, it does not provide a mechanism for the matter which makes up a human brain to have free will. Both deterministic and probabilistic models follow rules which do not leave any room for free will. A probabilistic model provides less predictability for individual particles but does not allow the particles to choose how they will act or have a mind of their own.

Many see the principles of quantum mechanics as giving human beings a special role. By measuring the position of the electron around an atom, humans can calculate the probability of finding the electron at any position and thereby confirm the probabilistic interpretation of the wave function (which is known as the Copenhagen interpretation). As described by Professor Weinberg:

> *Where human beings had no special status in Newtonian physics, in the Copenhagen interpretation of quantum mechanics, humans play an essential role in giving meaning to the wave function by the act of measurement.*

In my opinion, this statement by Professor Weinberg implies there is something special about humans that no other matter on earth possesses. Rather than quantum mechanics being able to explain human consciousness, it is human consciousness that exists outside of matter and gives a special interpretation to a wave function. Without human consciousness and free will, the wave functions of the universe, including the wave functions inside human brains, would presumably continue on their ethereal, nonreal (or real), and very deterministic existence. The fact that human consciousness and free will use the matter in the brain and the human body as tools to make measurements does not imply that human consciousness and free will are caused by the matter in the brain.

There is, however, another reason for questioning the "reality" of the wave function. The "reality" of the wave function bumps against one very fundamental aspect of science. Scientific evidence (or identifying what is "real") is based on experiments that measure the property predicted by the model. No one, however, has ever been able to directly measure the value of a wave at a point in space (although there are experiments that provide evidence for the wave nature of electrons). As Professor Weinberg explains, it is not a wave *of* anything. Every time the electron is measured, the wave function collapses (or the electron wave is rebundled) and the electron is found at a certain position with particle characteristics such as charge, mass, and spin. If no one has ever seen an electron in the form of a wave function around an atom, how can we be sure we know what it is doing (or if it is even there) when we are not looking at it?

If something, such as the wave function, explains how the universe works but cannot be measured, is it "real?" As attested by several comments (quoted previously), in countless experiments since the 1920s, all experiments have been consistent with the theory of quantum mechanics. But, the wave function, which forms the basis of quantum mechanics, has never been directly observed. On the other hand, remember the observation of Professor Hawking that the existence of virtual particles is inferred by the effect they have, not by direct measurement.

If all of this is confusing, do not feel alone. Although physicists know how to use quantum mechanics to get predictable results, none of them actually "understands" how it "really" works. Professor Weinberg tells the story of a promising physics graduate student who dropped physics after making the mistake of trying "to understand quantum mechanics."[80] Then, in reflecting on the

---

80. Weinberg (1992), pg. 84.

reality of the wave function and the question as to how probabilities arise from a deterministic wave function, he concludes:

> *But I admit to some discomfort in working all my life in a theoretical framework that no one fully understands.*[81]

Does this statement mean that quantum mechanics is just a lucky mathematical model without having any basis in reality? Is it just a coincidence that quantum mechanics "explains" how the universe works, at least when general relativity is not taken into account in cosmological situations? Since all experiments rely on experimental apparatus, is there really a distinction between what is "directly observed" and what is just inferred about the nature of matter? What kind of evidence do we need before we believe something is true? Remember, as discussed in Chapter 1, the goal of science is to develop hypotheses and theories that are consistent with the evidence. Choosing to believe what is "real" or "true" is outside the realm of science. And yet as human beings, we do choose which evidence is valid and reliable enough that we can base on it what we believe and how we should live.

This discussion is important because I have used a similar approach in providing evidence that leads to the conclusion supernatural souls exist. Although supernatural souls cannot be directly detected, their existence is inferred by the evidence that all natural phenomena follow the rules of science and the evidence that humans have free will. Supernatural souls explain free will, while no theory that relies on matter that interacts according to the laws of nature can explain free will.

Chapter Conclusions

1. Classical (Newtonian) physics is based on a deterministic model of the universe.

2. Quantum physics is based on a probabilistic model of the universe, although the wave function evolves in a deterministic fashion.

---

81. Weinberg (1992), pg. 85.

3. Neither a deterministic nor a probabilistic model allows the matter inside a human skull to have a free will or a mind of its own.

4. Particles of light (photons) and subatomic particles such as electrons, protons, and neutrons cannot make free will choices. Although they sometimes exhibit characteristics of waves and sometimes exhibit characteristics of particles, it is the nature of the experiment that determines which characteristic is exhibited and not the "choice" of the particle or wave.

5. There is no model or explanation in which particles or waves that do not have the ability to make free will choices can somehow be arranged to be able to make free will choices.

6. Quantum mechanics does not provide an explanation for the source of free will choices although it might provide a description of the mechanism of how free will choices and decisions are implemented by the body.

7. It is the ability to make free choices that defines free will. The inability to model a complex interaction is not a description of free will.

8. People, including scientists, base their beliefs on evidence. There is overwhelming evidence that all natural phenomena (including the atoms and molecules that make up human brains) are subject to the laws of physics. This provides evidence that the atoms and molecules that make up human brains cannot be the source of human free will.

9. The scientific models, such as string theory and super-symmetry that are being studied and may one day replace the currently held standard model, do not include any types of particles, fields, or wave func-

tions that would have minds of their own or be able to explain free will.

10. A belief that matter can have a "mind of its own" and decide to contradict the laws of physics would not be consistent with decades of scientific evidence.

11. Scientific theories such as the wave function in quantum mechanics are based on conceptual models and inferences from experimental results rather than direct observation. This does not prevent scientists from believing that they provide a useful explanation as to how the material universe operates.

"It would appear that it [the brain] is the sort of machine a 'ghost' could operate."

—*Sir John C. Eccles, 1963 Winner of the Nobel Prize for his Research on the Transmission of Nerve Impulses*

Chapter Six

THE SOUL-BRAIN INTERFACE

René Descartes, a philosopher of the 1600s, is known as the Father of Rationalism because he believed that truth could be arrived at by logical reasoning. He reasoned that his ability to doubt his own existence provided evidence that there must be a "thinking thing" that was doing the doubting. This concept was encapsulated in his famous saying: "I think, therefore I am." He believed that at the most basic level, he was a "thinking thing." He concluded that humans have a dual nature: bodies made out of a "material substance" that is extended in time and space (*res extensa* in Latin) and a supernatural soul that is the "thinking thing" (*res cogitans* in Latin). He theorized that a supernatural soul might interact with the brain in the pineal gland, a small gland in the center of the brain. This is known as Cartesian dualism or Cartesian interactionism.

It is now well accepted by the scientific community that the brain is the part of the human body that "thinks" and directs most of the activities of the rest of the body. However, Mr. Descartes' suggestion about a soul-brain interface in the pineal gland has been rejected by most scientists because they think there is no scientific model that explains how an immaterial soul could interact with a material brain.

Some scientists and philosophers have attempted to develop models of the mind which rely on the human brain being the "thinking thing" and having consciousness. However, a human brain is made out of atoms and molecules that interact based on the laws of physics. This leaves no room for free will and implies that free will cannot be explained as a biological mechanism. Thus, it is logical to conclude that we must look outside the material brain for the source of free will. We must look to a supernatural agent that can be free to choose and that is not bound by the forces of nature and nurture. This supernatural agency is usually referred to as a soul.

Mr. Descartes' soul-brain theory was developed during the period when Isaac Newton discovered the laws of motion and gravitation. Experiments indicated that the motion of each material object is caused by its interaction with other material objects such that there is always conservation of energy and momentum. Thus the motion of each material object was visualized to be like that of a billiard ball, bouncing into and bouncing off of other billiard balls. Even the movement of the particles of the brain were imagined to be like some intricate mechanical calculating machine that would control the action of the body in some mysterious way. At that time, electricity, chemistry, and quantum physics were unknown. The concept of brain neurons and synapses working via electrochemical signals was several hundred years in the future.

With the Newtonian concepts of the material world, it was difficult if not impossible to imagine how a supernatural, non-material soul could "push" or be "pushed by" the particles of the brain. Like our modern version of ethereal ghosts that can pass through physical barriers, a supernatural soul would likewise pass through the physical matter of the brain without being able to interact with it. Since the existence of a supernatural soul has never been detected by scientific instruments or by scientific measurements, many scientists consider belief in the soul to contradict science. However, such belief by scientists is a misunderstanding of the nature of science. Science does not claim there is no supernatural world. Science only claims to explain how the material world functions; the supernatural world is not subject to direct scientific investigation or verification. The fact that souls have not been detected by scientific instruments does not make souls unscientific. On the contrary, supernatural souls are the most logical and scientific explanation of human free will. Atoms and molecules having free will and "minds of their own" would not be consistent with decades of scientific evidence. Possible mechanisms for a soul-brain interface and interaction are described below.

THE NATURE OF MATTER HAS CHANGED

Starting in about 1900, scientists discovered a radical change in the understanding of the nature of matter. With the formulation of the theory of quantum mechanics, matter was seen to have a dual nature. Experiments indicated that matter can exhibit both a particle and a wave nature. Matter also became more "fuzzy" as the Heisenberg uncertainty principle declared that there is a limit to how closely it is possible to simultaneously measure the velocity and position of a particle. Edwin Schrödinger developed a wave equation which described the existence of electrons as waves that are spread out around atoms. Evidence was also discovered that an electron can jump from a high energy state in the atom to a lower-energy state, with the energy difference showing up in the form of a photon of light that is emitted from the atom. Conversely, an electron around an atom can absorb energy in the form of a photon of light which moves the electron up to a higher energy state.

In his famous equation $E = MC^2$, Albert Einstein declared that matter (mass) and energy are equivalent. They are just different forms of the same natural phenomenon. Suddenly, matter was no longer made up of the tiny hard particles envisioned in the mechanical physics of Isaac Newton. Matter is no longer considered to be "solid." First of all, most of an atom is empty space. The atomic nucleus makes up 99 percent of the weight of an atom but only a small percentage of the volume. The electrons exist as waves in probability shells around the nucleus. The volume of the atom, as defined by the outermost electrons, is filled mostly with empty space. The reason a person's hand cannot go through a table is not because both are "solid." Rather it is because the negatively charged electrons in the outer shells of the atoms that make up the hand are repelled by the negatively charged electrons in the outer shells of the atoms that make up the table.

Secondly, even the components of the atom (the protons, the neutrons, and the electrons) are not solid. The components of atoms are now envisioned to be bundles or packets of energy fields.[1] What we detect as "particles" are only manifestations of those fields. In other words, we only perceive the bundles of the energy fields as particles. With matter envisioned as packets of energy fields rather than as "solid" particles, the brain no longer has to be imagined as an intricate mechanical clock-like mechanism. It is now possible to hypothesize that a supernatural soul can interact with small bits of matter in the brain by creating energy that can change the energy state of matter in the brain.

From a scientific standpoint, the above hypothesis is a reasonable explanation for the transmission of information from one entity to another. Note that based on scientific principles, information cannot be transmitted without the concurrent transmittal of energy. For the example considered in this book, the information that is transmitted is the information associated with directing the brain to perform willful acts. Since scientific evidence indicates that a material brain cannot make willful decisions on its own, it is logical to conclude that the information is transmitted from a supernatural soul to the brain.

POSSIBLE MODES OF TRANSMITTING INFORMATION

There are various mechanisms that might be used to transmit information between the soul and the brain. Although there is not yet any definitive evidence as to how such information might be exchanged, there are at least some plausible theories that are consistent with known scientific principles.

---

1. See the chapter "Quantum Mechanics."

One theory is based on the soul interacting with atomic particles in the neurons of the brain. The current neurological model envisions the brain as a computer-like information-processing and information-storing organ that is made up of billions of brain cells called neurons. Each brain cell has thousands of tentacles called dendrites that connect to other brain cells in an incredibly complex arrangement. The dendrites connect to the body of the brain cell (called the soma), and the soma is connected to the axon of the brain cell. The many branches of an axon from one brain cell connect to the dendrites of other brain cells. At the point of connection between an axon of one brain cell and the dendrite of another brain cell, there is a gap called a synapse. In *The Mind and The Brain,* Dr. Jeffrey Schwartz indicates there are estimated to be one hundred billion neurons in a human brain with 100 trillion to 1,000 trillion synaptic connections.[2] The following diagram shows a typical brain cell with dendrites, axons, and synapses.

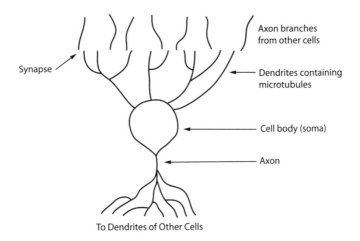

Figure 6.1

Simplified sketch of a brain cell (neuron)

(Not to scale and not all dendrites and axon branches are shown)

2. Schwartz (2002), pg. 111.

An electrochemical signal is transmitted from the axon of one brain cell to the dendrites of other brain cells by the movement of fluid across the synaptic gaps. The synaptic fluid is made up of neurotransmitters.[3] The signal (in the form of electrically charged particles) then moves from the dendrites to the body of the brain cell (the soma). The signals from thousands of dendrites are accumulated in the soma, and then another signal is sent to the axon where it is transmitted across other synaptic gaps to the dendrites of other brain cells.

Some scientists theorize that the signals transmitted across the synaptic gaps of the brain and from somas to axons are comparable to the signals of a digital computer. A digital computer functions based on binary computations using "1's" and "0's." Logic gates in the computer switch between the "on" or "off" mode to perform the binary computations. The "on" mode simulates "1's" and the "off" mode simulates "0's." Likewise, it is theorized that these "on" and "off" functions can be simulated by synapses that either "fire" by transmitting fluid or do not "fire" by not transmitting fluid. Also, the neuron itself can "fire" and send a signal to the axon when it accumulates enough signals from the dendrites. Thus the "firing" of the synaptic gaps and the "firing" of the neurons might simulate the switching action of computer logic gates. Other scientists have theorized that the brain functions in a continuous ("analog") mode rather than digitally:

> *The brain's activities are known to use analog, electrochemical systems to achieve thought. The neurons communicate among themselves, not with digital 1's and 0's, but with voltage spikes that stimulate the production of chemical transmitters. As more spikes arrive at a synapse in a given amount of time, the synapse releases more neurotransmitters.*[4]

---

3. In general, drugs affect brain activity by affecting the chemistry of neurotransmitters.

4. Johnson (1988), pg. 173.

The structure and functioning of the brain neurons and synapses are described in great detail in the 1994 book *How the Self
Controls Its Brain* by Nobel prize-winning brain researcher Sir
John Eccles. In his book, Professor Eccles avoided the use of the
term "soul," because, I think, he believed it to be a controversial
and emotionally charged word. He used instead the more neutral
word "mind" to refer to the decision-making entity of the self.
However, he was very clearly a "dualist," believing that there is a
dual supernatural and material nature of human beings. In 1952,
Sir Eccles responded to the challenge of philosophy professor
Gilbert Ryle who, in his 1949 book *The Concept of Mind*, rejected
Cartesian dualism and Cartesian interactionism. Professor Ryle
derisively referred to dualism as the failed concept of "the ghost
in the machine." Professor Eccles described how the nature of
matter has changed such that it is now possible for the brain to
be a "machine" that could be operated by a "ghost:"

> The neurophysiologist, Sir John Eccles, in his Waynflete
> Lectures of 1952 at Magdalen College, Oxford, presented his
> modern-day version of Cartesian interactionism and closed his
> peroration with what he must have supposed was a graceful
> bow to the Waynflete Professor of Philosophy [Gilbert Ryle]:
> "If one uses the expressive terminology of Ryle, the 'ghost' oper-
> ates a 'machine,' not of ropes and pulleys, valves and pipes,
> but of microscopic spatio-temporal patterns of activity in the
> neuronal net. . . . It would appear that it is the sort of machine
> a 'ghost' could operate." . . . Eccles, J. C. (1953) The Neuro-
> physiological Basis of Mind. Oxford: Oxford University Press.
> p. 285.[5]

As Sir Eccles explains in this excerpt, the soul (the "ghost")
does not have to operate a Newtonian brain that is like a machine
of "ropes and pulleys, valves and pipes." Rather, the soul interacts
with the brain by changing the activity in the synapses of the

---

5. Dennett (1984), pp. 6–7.

brain neurons (that is, the "microscopic spatio-temporal patterns of activity in the neuronal net"). He theorizes that the mind interacts with the brain using the principles of quantum mechanics in the synaptic gaps of the brain cells:

> . . . *the hypothesis of mind-brain interaction is explained by [Professor] Beck using quantum physics without infringing the conservation laws of physics (Chapter 9).*[6]

In chapter 9 of his book, Professor Eccles provides the detailed quantum mechanical analysis of synaptic gaps developed by Professor Beck. The model that is developed relies on quantum mechanics which does not allow for predicting the firing of individual brain synapses on a deterministic basis. Instead, the quantum mechanical model predicts the synaptic gaps would fire in a probabilistic fashion. The firing of brain synapses based on probabilistic principles would most likely increase the complexity and difficulty in understanding how the brain functions. Professor Eccles also believed that the brain might function using some type of quantum coherence.[7]

Professor Eccles theorized that the supernatural soul, in the form of psychons,[8] interacts with the dendrons[9] of the brain. Apparently Professor Eccles considered each psychon to be like a "little piece of soul" which is associated with a specific dendron. Professor Eccles and Professor Beck theorized that the movement of neurotransmitters across the synaptic gap might be controlled by the quantum mechanical energy state of a relatively few molecules in the synaptic junction. Professor Eccles recognized that the molecules in the synaptic junctions of the brain cannot

---

6. Eccles (1994), p. xi. The conservation law referred to in the above excerpt is the requirement that the probability of firing plus the probability of not firing must equal one.

7. Quantum coherence is described in more detail later in this chapter and in the chapter "Quantum Mechanics."

8. "Psyche" is the Greek word for soul or mind.

9. A dendron is a group of dendrites.

decide on their own to change their own energy states. Thus, he believed that there has to be a mental force (a supernatural soul) that initiates human free will decisions and controls the brain.

Under the above scenario, the soul (mind) might implement a free will decision by changing the energy state of the molecules in the synaptic junctions of the brain. This change in the energy state would increase the probability that the synaptic junction would "fire" by releasing neurotransmitters across the junction. Thus the mechanical pushing and pulling of material objects envisioned at the time of René Descartes is not required. Rather, a soul that can create infinitesimal amounts of energy needed to affect the energy state of a few electrons in certain synaptic gaps could change the probability that they will fire. Thus, they could theoretically control the "on/off" switching of the brain cells, thereby controlling the decision-making process of the brain.

In his 1984 book *Minds, Brains, and Science*, Professor John Searle explains that neither classical nor quantum physics provides an explanation as to how a being made up only of matter can have free will. Professor Eccles believes that he has solved the "free will" problem of Professor Searle:

> Our[10] hypothesis offers a natural explanation for voluntary movements caused by mental intentions without violating physical conservation laws. It has been shown experimentally that intentions activate the cerebral cortex without infringing the conservation laws of physics (Beck and Eccles 1992 and chapter 9 [of Eccles 1994]). So Searle (1984, p. 99) . . . can continue with a good conscience in voluntary, free, intentional actions![11]

Professor Eccles appears to be saying that a mind (or soul) could affect the quantum mechanical state of the electrons in the synaptic junction of brain neurons without expending any energy.

---

10. "Our" refers to Professor Eccles and Professor Beck.

11. Eccles (1994), pp. 169–170.

However, in an article entitled "Mind-Brain Interaction and Violation of Physical Laws" in the 1999 book *The Volitional Brain: Towards a Neuroscience of Free Will*, Professor David L. Wilson[12] describes how his calculations show that there are errors in the calculations of Sir John Eccles and Professor Beck and that the amount of energy associated with the firing of the synapses of the brain exceeds the level of energy that would be classified as quantum mechanical uncertainty. He points out that even if the energy level is below the quantum mechanical uncertainty level, it would still require *some* energy to change the probability that any of the synaptic junctions of brain neurons will fire. Also, he claims that the specific mechanism envisioned by Professor Eccles is unlikely, considering current knowledge about brain neurology. Nevertheless, Professor Wilson presents a number of alternative mechanisms for affecting the firing of brain synapses, all of which require energy levels that are greater than quantum mechanical uncertainty levels. These alternative mechanisms include opening sodium channels, altering voltage gradients, synaptic transmission involving the presynaptic terminal, and synaptic transmission involving the postsynaptic membrane. Thus, all that is needed is a source of energy to accomplish, at least conceptually, what was envisioned by Professor Eccles, even if the specific mechanism is different. I agree with Professor Wilson that energy[13] needed to affect the firing of brain synapses requires a violation of natural laws (and thus would be a supernatural phenomenon). I agree with the following observation of Professor Wilson:

> *For those who believe that free will must be totally independent and free of physical causes, the idea that physical laws must be violated should not be taken as a negative but almost as an expectation, especially to the extent that physical laws*

---

12. Professor Wilson is a professor of biology at the University of Miami. He has a B.S. in physics from the University of Maryland and a Ph.D. in biophysics from the University of Chicago.

13. This is referring to energy from a source other than the brain.

appear to specify a universe that is either determined or ran-domly probabilistic. If a nonphysical mind exists, the research project for the next century should be to explore the impact of such nonphysical influences—where in the brain does such influence occur and what laws are broken?[14]

Although Professor Wilson estimates that the energy needed to affect brain synapses exceeds the level that would be considered to be quantum mechanical uncertainty, Dr. Schwartz, whose book was reviewed by physicist Henry Stapp, opines that quantum mechanical effects do need to be considered in the operation of brain synapses:

Applying quantum theory to the brain means recognizing that the behaviors of atoms and subatomic particles that constitute the brain, in particular the behavior of ions[15] whose movements create electrical signals along axons and of neurotransmitters that are released into synapses, are all described by Schrödinger wave equations. Thanks to superpositions of possibilities, calcium ions might or might not diffuse to sites that trigger the emptying of synaptic vesicles, and thus a drop of neurotransmitter might or might not be released.[16]

Regardless whether or not the energy needed to affect brain synapses is below quantum mechanical uncertainty, I agree with Professor Wilson that there would have to be some energy expended in order to transmit "free will" information back and forth from the soul to the brain. It is instructive to estimate the relative amount of energy that might be involved in transmitting a willful decision from the soul to the brain. The amount of energy required can be estimated in comparison to the equivalent energy of the entire mass of a human body using Einstein's equivalence formula $E = MC^2$. An estimate of this quantity of energy is based on the combination of the following factors:

---

14. Libet (1999), pg. 196.

15. Ions are particles that have either a positive or a negative electrical charge.

16. Schwartz (2002), pg. 284.

- Item 1: The weight of the entire human body compared to the weight of the brain
- Item 2: The weight of the brain compared to the weight of the affected molecules in the synaptic junctions of the brain
- Item 3: The weight of the affected molecules compared to the weight of the affected electrons in the molecules whose energy states are changed
- Item 4: The weight of the affected electrons compared to the equivalent energy represented by the change in energy state of the electrons

The order of magnitude of the values of the above items can be estimated as follows:

- Item 1:   50 times
- Item 2:   1,000,000 times
- Item 3:   1,000,000,000,000 times[17]
- Item 4:   100,000 times

Combining the above factors (by multiplying them together) results in an estimate that the weight of the human body is 5 trillion trillion times greater (in equivalent energy) than the energy needed to affect the energy state of molecules in the synaptic junctions of the brain to cause them to fire. From this, it should be fairly obvious that it would be impossible to directly detect a soul "leaving" the body by weighing the body at the time of death. First of all, no scale is accurate enough to detect such a small amount of weight. Secondly, the energy equivalence of the weight of the air leaving the body as each breath is exhaled or even the

---

17. In a footnote on page 110 in his 1968 book *Scientific Indeterminism and Human Freedom*, Professor Margenau notes that "A synaptic knob within the cerebral cortex weights 5×10^{-13} gm, i.e. 10^{14} times as much as an electron. J.C. Eccles, (*The Neurophysiological Basis of Mind*, Oxford, 1953) considers the possibility that particles within these knobs, free under physical indeterminacy, may have their behavior influenced by the mind."

energy given off from the body as heat by evaporation and radiation is likely to be many billions or trillions of times greater than the energy needed to fire the synaptic junctions in the brain.

Likewise, detecting a soul by somehow detecting its energy would also be very difficult due to limitations in the accuracy of scientific measuring techniques, even if, theoretically, there were a way to independently detect the presence of a soul. The previous estimate is intended to give only a rough order of magnitude and is not meant to be scientifically rigorous. However, even if it is wrong by a factor of one trillion times, it gives an idea as to how subtle is the effect that is needed to direct the activities of the brain. There are many reasons why it will be difficult to directly detect the action of a supernatural soul as it affects a brain:

- The effect is so subtle compared to the surrounding brain matter.
- The molecules affected are buried inside the brain of a living person, making it very difficult to determine which and when any of the trillions of synapatic junctions are firing.
- Even if we could somehow monitor whether or not specific synapses are firing, it would be difficult to determine which synapses are firing due to the effect of the soul, which are firing due to the supplemental action of the brain to implement any free will decision, and which are firing due to some other brain function.
- The soul might be able to absorb communications from the brain to receive information on how the body "feels." The energy flowing from the body to the soul might offset some of the energy flowing from the soul to the body and thus reduce the net amount of energy flowing between the soul and brain.

There are still many scientific and "engineering" problems that need to be resolved to develop a good "soul-brain interface" model. These include:

- Determining the type of mechanism or mechanisms used by the soul to affect the energy state of the molecules in the synaptic junctions of the brain, causing them to fire.
- Determining the type of mechanism or mechanisms used by the soul to receive information from the brain.
- Determining how the soul stays with the brain of one person as that person moves through time and space and how the soul identifies within which synaptic junctions to interact.

I propose the following possible mechanism that the soul might use to affect the energy state of the molecules in the synaptic junctions of the brain. This mechanism is one in which the soul creates photons of light near the appropriate molecules. These photons of light could cause electrons in the molecules to change their energy state, thus changing the probability that the synapse will fire. This theory requires that a supernatural soul be able to create energy in the form of light. This power to create energy would be one of the God-like powers that humans possess even though the amount of light created by humans would be miniscule compared to the equivalent amount of light energy present at the instant of the big bang beginning of the universe. One interesting aspect about this soul-light switching theory is that there are similarities between it and ongoing research to develop a light-switching digital computer that would increase the computational speed of computers.

One other possible model is to imagine each dendrite as a tiny antenna that uses the electrically charged ions moving through the string-like dendrites of the brain neurons to radiate nearly imperceptible radio signals that are picked up by the soul. The soul, in turn, would communicate with the brain by creating radio signals[18] that affect the electrically charged particles in the den-

---

18. Remember that radio signals are another form of electromagnetic radiation, which is also known as "light."

drites, causing current to flow, which results in the firing of brain synapses. This mechanism is consistent with one of the firing modes described by Professor Wilson, noted previously. This mechanism is also consistent with Maxwell's equations which define the interrelationship between moving charged particles and electromagnetic radiation, which is also known as light.

The answer as to how the soul stays with the brain as it moves through time and space might be that the soul knows the three-dimensional arrangement and connections of the brain and continually sends and receives test signals to verify the brain identity. Another possibility might be that the soul uses microwave vibrational patterns[19] to detect the genetic code found in the DNA in each cell of the human body. This theory immediately raises the question as to how the souls of identical twins can determine which brain to control. There might be subtle differences in the billions of genetic combinations making up the genetic codes of twins that would allow the souls to discriminate between the brains of twins.

OTHER POTENTIAL SOUL/BRAIN MECHANISMS

As time goes on, there will likely be other models of the soul-brain interaction. Undoubtedly, the interaction of the soul with the brain is very complex and likely involves some type of mechanism we have no way of even imagining at this time. Several other potential soul-brain mechanisms are described briefly below:
- Microtubules/cytoskeletal action
- Quantum coherence and probability fields
- Microwave interaction

MICROTUBULES/CYTOSKELETAL ACTION

The concept of using quantum mechanical or biochemical action in the microtubules and cytoskeletal parts of the brain to

---

19. See the information later in this chapter concerning the interaction of the brain with microwave radiation.

"explain" free will is developed in the 1994 book *Shadows of the Mind: A Search for the Missing Science of Consciousness ("Shadows")* by Professor Roger Penrose. This concept is similar to that proposed by John Eccles (as described previously) in that it relies on an unknown quantum mechanical mechanism in the brain neurons. However, unlike John Eccles, Professor Penrose believes that the quantum mechanical mechanism occurs in small biological tubes ("microtubules") located in the dendrites. (See Figure 6.1.) More importantly, Professor Penrose does not believe in supernatural souls. Thus, he believes that the microtubule/cytoskeletal action could be more than just the *mechanism* used to transmit a free will decision. He believes that it might be the *source* of free will.

Professor Penrose believes there has to be some noncomputational aspect of quantum mechanics that has not yet been discovered. He is looking for a noncomputational mechanism because, as described in the chapter "Math," he knows there are noncomputational aspects of human thinking that cannot be explained by a system that follows computational rules. Professor Penrose believes that the explanation of free will and human consciousness might be found in some type of unknown noncomputational aspect of the quantum mechanical wave function collapse,[20] which he refers to as the objective reduction (OR). He considers the OR action to be the interplay at the "borderline" between the quantum mechanical world and the classical world. Professor Penrose envisions this OR action to take place in the microtubules that are located in the cytoskeletal structure of the brain cells. This cytoskeletal activity would then influence the firing of the neurons which is the computer-like switching that is described above.

Professor Penrose's theory also relies on some type of quantum coherence. Quantum coherence means that subatomic particles in various parts of the brain act in unison. This is

---

20. The collapse of the wave function is discussed in the chapter "Quantum Mechanics."

needed, according to his theory, to give humans the feeling of "now" and to tie perceptions together. Professor Penrose summarizes his theory as follows:

> I am suggesting that a physical noncomputability—needed for an explanation of the noncomputability in our conscious actions—enters at this [quantum-classical borderline] level. . . . I argue that whereas neuron signals may well behave as classically determinate events, the synaptic connections between neurons are controlled at a deeper level, where it is to be expected that there is important physical activity at the quantum-classical borderline. The specific proposals I am making require that there be large-scale quantum-coherent behaviour (in accordance with proposals that have been put forward by Frohlich) occurring within the microtubules in the cytoskeletons of neurons. The suggestion is that this quantum activity ought to be noncomputationally linked to a computation-like action that has been argued by Hameroff and his colleagues to be taking place along microtubules. . . . I feel that it is important that any dedicated reader, wishing to comprehend how such a strange phenomenon as the mind can be understood in terms of a material physical world, should gain some significant appreciation of how strange indeed are the rules that must **actually** govern that "material" of our physical world.[21] [Emphasis in the original]

Note that in the last sentence of the excerpt Professor Penrose describes that matter must act according to rules, even if they are "strange," noncomputational rules. This is commented on later in this chapter. Professor Penrose further describes his theory:

> On the view that I am tentatively putting forward, consciousness would be some manifestation of this quantum-entangled internal cytoskeletal state and of its involvement in the interplay (OR) between quantum and classical levels of

---

21. Penrose (1994), p. vi.

*activity. The computer-like classically interconnected system of neurons would be continually influenced by this cytoskeletal activity, as the manifestation of whatever it is that we refer to as "free will." The role of neurons, in this picture, is perhaps more like a **magnifying device** in which the smaller-scale cytoskeletal action is transferred to something which can influence other organs of the body—such as muscles. Accordingly, the neuron level of description that provides the currently fashionable picture of the brain and mind is a mere **shadow** of the deeper level of cytoskeletal action—and it is at this deeper level where we must seek the physical basis of **mind!** There is admittedly speculation involved in this picture, but it is not out of line with our current scientific understanding.*[22] *[Emphasis in the original]*

Professor Penrose's description of neuronal action as being a magnifying device is similar to the above analysis that shows how an extremely small amount of energy could affect the computer-like switching of the brain, resulting in the expenditure of much larger quantities of energy by the body. I agree that the cytoskeletal action might be able to explain the *manifestation* of human free will and consciousness, but it cannot be the *source* of human free will and consciousness. It is not completely clear whether Professor Penrose considers the cytoskeletal action or the computer-like activity of the neurons as the manifestation of free will. I would only comment that he has not, even in concept or theory, identified any physical process of the cytoskeletal action that would explain how matter could have a mind of its own to produce free will. He recognizes that the cytoskeletal action must follow rules, even if they are strange, noncomputational rules. Thus the cytoskeletal action may be the manifestation of free will but not its source.

As described in the following excerpt, Professor Penrose recognizes that current physics does not allow matter to perform

---

22. Penrose (1994), p. 376.

the noncomputational actions he feels are needed to account for consciousness.[23]

> *Does present-day physics allow for the possibility of an action that is in principle impossible to simulate on a computer? The answer is not completely clear to me, if we are asking for a mathematically rigorous statement. Rather less is known than one would like, in the way of precise mathematical theorems, on this issue. However, my own strong opinion is that such noncomputational action would have to be found in an area of physics that lies **outside** the presently known physical laws.[24]* [Emphasis in the original]

I agree with Professor Penrose that the currently understood laws of physics cannot simulate noncomputational rules although it is apparently difficult to prove this mathematically. Examples of noncomputational rules can be found in the chapter "Math." However, even if matter can perform noncomputational actions, it must still function according to rules. Matter cannot be the source of free will as long as it must act according to rules, even if they are noncomputational rules. Professor Penrose is searching for physical processes governed by noncomputational rules that lie "outside the presently known physical laws" but still within the realm of science which operates according to rules. Noncomputational rules do not mean that the matter would be able to do anything it wants, only that the rules are not computational rules, such as those followed by a computer. Anything that acts according to rules, even noncomputational rules, does not have the freedom to have a mind of its own and have free will. Thus, matter acting according to noncomputational rules might be the manifestation of free will but it cannot be the source of free will.

---

23. See the chapter "Math" for further discussion on the ability of humans to perform noncomputational actions and to solve problems noncomputationally.

24. Penrose (1994), p. 15.

MICROWAVES

The ability of biological cells, including those in human brains, to react to microwaves is also mentioned in *Shadows*. Professor Penrose describes how the biological vibration sets up an interaction with the microwave radiation in a mode of quantum coherence. In this case, the microwave radiation at 10,000 megahertz (10^{11} hertz) is the mechanism for the quantum coherence. Professor Penrose explains:

> . . . *a puzzling phenomenon . . . had been observed in biological membranes as far back as 1938, and Frohlich was led to propose, in 1968 (employing a concept due to my brother Oliver Penrose and Lars Onsager (1956)—as I learnt to my surprise when looking into these matters), that there should be vibrational effects within active cells, which would resonate with microwave electromagnetic radiation, at 10^{11} Hz, as a result of a biological quantum coherence phenomenon. Instead of needing a low temperature, the effects arise from the existence of a large energy of metabolic drive. There is now some respectable observational evidence, in many biological systems, for precisely the kind of effect that Frohlich had predicted in 1968.[25]*

Note that microwave radiation is the same phenomenon as radio waves and visible light but at a higher frequency. Microwave radiation operates based on well-defined laws of physics. It does not have a mind of its own and cannot be the source of free will. It might, however, be the mechanism by which free will decisions are implemented in the brain.

QUANTUM COHERENCE AND PROBABILITY FIELDS

As described above, Professor Eccles theorizes that quantum coherence might be involved in brain activity and Professor Penrose relies on large-scale quantum coherence for his theory of consciousness. In her 1990 book *The Quantum Self*, Professor

---

25. Penrose (1994), p. 352.

Danah Zohar theorizes that consciousness is made possible in the brain by a type of quantum coherence called a Bose-Einstein condensate. This is similar to the quantum coherence concept referenced by Professor Penrose.[26] Quantum coherence might be a manifestation of free will and it might be the mechanism by which free will decisions are implemented, but it has no ability to act on its own and is thus not the source of free will.

In his 1984 book *The Miracle of Existence*, Professor Henry Margenau envisions the human mind to be a "field in the accepted physical sense of the term." But he considers it to be a "nonmaterial field; its closest analogue is perhaps a probability field. . . . And so far as present evidence goes, it is not an energy field in any physical sense . . . ."[27] I have a similar comment concerning Professor Margenau's theory: Any entity that is subject to the laws of physics, even if it a nonmaterial probability field, cannot be the source of free will.

DECISION SEQUENCES IN THE BRAIN

In his 1991 book *The Brain Has a Mind of Its Own*, neurologist Richard Restak, M.D. reports on brain research by Benjamin Libet,[28] a neurophysiologist at the School of Medicine of the University of California at San Francisco. Professor Libet found that there is a "readiness potential" that is electrically measurable in the brain about one-third of a second before an experimental subject indicates he or she has decided to voluntarily move his or her finger. Dr. Restak concluded that this evidence indicates "each brain has a mind of its own."[29] In my opinion, this does not provide evidence that the brain can make a free will decision on

---

26. See the discussion on quantum coherence in the chapter "Quantum Mechanics."

27. Margenau (1984), p. 97.

28. The research is also described in Libet (1999).

29. Restak (1991), pg. 50.

its own. There are several interpretations of the results that support the understanding that humans have free will that affects the brain and not the other way around. Under one interpretation, the results indicate that a person has a difficult time in identifying exactly when in the process of getting ready to move a finger that the actual decision to move has been made.[30] For example, the readiness potential might indicate the person is thinking about moving the finger or is getting ready to decide to move the finger. The decision to act would then occur after the readiness potential is registered.

In the 1999 book *The Volitional Brain*, edited by Benjamin Libet, Gilberto Gomes explains that Sir John Eccles theorized the brain is continually producing random readiness potentials that the immaterial "mind" makes use of to initiate a free will decision or action. With this understanding, the readiness potential can appear first followed by the decision to act. This explains how the readiness potential can occur before the time the subject reports that a decision has been made.

Another interpretation makes use of the fact that a free will decision can be a decision to not do something instead of a decision to do something. In his research, Professor Libet found that the subject can cancel a decision to act "between the awareness of intent [to move the finger] and the action itself."[31] Professor Libet concluded that the results support the existence of human free will (or as some have called it "free won't"). With this understanding, the results can be interpreted as follows. The subject makes multiple decisions to get ready to lift the finger (hence the readiness potential) but chooses to cancel the decision to act (or to not follow through on the action) for each sequence except for the last sequence. For the last sequence, the subject chooses to not cancel the action. Thus, there is the readiness potential

---

30. See the paper by Gilberto Gomes in Libet (1999).
31. Restak (1991), pg. 49. Also, see Libet (1999).

followed by the subject's reported intent to act followed by the action. The concept of "free won't" is consistent with the moral code of many religions that require the individual to reject actions based on bodily urges and desires (for example, the "Thou shall nots" in the Ten Commandments).

Finally, neither Dr. Restak nor Professor Libet provided an explanation as to how matter that must follow the laws of physics can on its own make a free will decision.

ADDITIONAL BRAIN RESEARCH

Additional brain research is described in the fascinating 2002 book *The Mind and The Brain: Neuroplasticity and The Power of Mental Force* by Jeffrey M. Schwartz, M.D. and Sharon Begley. They describe their experience in treating obsessive-compulsive disorder (OCD) by teaching the patient how to make free will decisions by using self-directed neuroplasticity ("mental force") to change the brain patterns that cause the OCD. To demonstrate how their research fits into current scientific thinking, the authors first describe the belief by some scientists that free will does not exist because free will cannot be explained as a natural phenomenon:

> *It is not merely that the will is not free, in the modern scientific view; not merely that it is constrained, a captive of material forces. It is, more radically, that the will, a manifestation of mind, does not even exist, because a mind independent of brain does not exist.[32]*

However, the fact that science cannot explain free will as a natural phenomenon does not imply that free will does not exist.

Dr. Schwartz is a psychiatrist who used brain imaging techniques such as positron emission tomography (PET) and functional magnetic resonance imaging (fMRI) in his research to map

---

32. Schwartz (2002), pg. 8.

the effect of various types of mental activity and mental force on various regions of the brain. He found that patients with OCD had brain patterns that were at the root of the problem:

> *The obsessions that besiege the patient seemed quite clearly to be caused by pathological, mechanical brain processes— mechanical in the sense that we can, with reasonable confidence, trace their origins and the brain pathways involved in their transmission. OCD's clear and discrete presentation of symptoms, and reasonably well-understood pathophysiology, suggested that the brain side of the equation could, with enough effort, be nailed down.*[33]

Dr. Schwartz realized that there was a part of the mind that could stand "outside and apart from the OCD symptoms, observing and reflecting insightfully on their sheer bizarreness."[34] He helped the patients to "learn a practical, self-directed approach to treatment that would give them the power to strengthen and utilize the healthy parts of their brain in order to resist their compulsions and quiet the anxieties and fears caused by their obsessions."[35] He provides evidence that "the mind can change the brain,"[36] through a process of neuroplasticity.

> *Neuroplasticity refers to the ability of neurons to forge new connections, to blaze new paths through the cortex [of the brain], even to assume new roles. In shorthand, neuroplasticity means rewiring of the brain.*[37]

Dr. Schwartz shows "that directed, willed mental activity can clearly and systematically alter brain function; that the exertion of willful effort generates a *physical force* that has the power to change how the brain works and even its physical structure."[38]

---

33. Schwartz (2002), pg. 12.

34. Schwartz (2002), pg. 13.

35. Schwartz (2002), pg. 13.

36. Schwartz (2002), pg. 15.

37. Schwartz (2002), pg. 15.

38. Schwartz (2002), pg. 18.

The mechanism he envisions for the mind-brain interaction is similar to that described above relative to interaction in the synaptic junctions of the brain as conceptualized by Sir John Eccles in the mid-1900s. As Dr. Schwartz explains:

> *I am using [the word]* **force** *to imply the ability to affect matter. The matter in question is the brain. Mental force affects the brain by altering the wave functions of the atoms that make up the brain's ions, neurotransmitters, and synaptic vesicles. By a direct action of mind, the brain is thus made to behave differently. It is in this sense of a direct action of mind on brain that I use the term* **mental force***. It remains, for now, a hypothetical entity.*[39] *[Emphasis in the original.]*

Dr. Schwartz also recounts the success that has been achieved by using cognitive therapy rather than drugs to treat and relieve depression.[40] Patients are taught how to change their brains and thinking patterns when they experience feelings of depression. There has likewise been success in using the mental force of the will to improve the condition of those who have suffered from strokes and Tourette's syndrome:

> *The will, it was becoming clear, has the power to change the brain—in OCD, in stroke, in Tourette's, and now in depression—by activating adaptive circuitry.*[41]

Dr. Schwartz appears to theorize that the ability of conscious human experimenters to choose which aspect of nature to investigate in quantum mechanical experiments somehow allows the will to be a natural phenomenon:

> *At its core, the new physics [of quantum mechanics] combined with the emerging neuroscience [of neuroplasticity] suggests that the natural world evolves through an interplay between two causal processes. The first includes the physical*

---

39. Schwartz (2002), pg. 318.
40. Schwartz (2002), pg. 245.
41. Schwartz (2002), pg. 250.

*processes we are all familiar with—electricity streaming,
gravity pulling. The second includes the contents of our consciousness, including volition.*[42]

I think that Dr. Schwartz's conclusion that volition and free
will are natural phenomena explained by quantum physics is a
misapplication of the concepts of quantum physics. Because natural phenomena adheres to the laws of physics, the ability of a
being with consciousness and free will to choose which aspect of
nature to investigate does not imply that the source of the free
will is a natural phenomenon as Dr. Schwartz appears to believe.
Rather, it implies that the source of free will is outside of nature.

In my opinion, the OCD treatments described above provide
significant evidence that a force outside the material brain can
choose to make an effort to go against the quantum mechanical
probabilities defining the OCD brain activity that would have
occurred absent the choice and the effort. There is nothing in
the theory of quantum mechanics that implies that atoms and
molecules (such as those in a human brain) can somehow choose
to change the probability of interacting with other atoms and
molecules. It seems clear that without this mental effort by the
patient, the brain patterns would have remained unchanged and
the OCD behavior would have continued.

One of the inherent problems in identifying a free will decision is determining what the brain would have done absent the
free will decision. In the case of people exhibiting OCD behavior,
the presumption would be that the OCD behavior would continue without some outside force affecting the brain patterns.
Thus, the ability of the will to change OCD behavior is evidence
that the will is changing brain patterns.

---

42. Schwartz (2002), pg. 19.

Chapter Conclusions

1. In the 1600s, René Descartes theorized that a supernatural soul interacts with the material brain in the pineal gland. Scientists, however, rejected this idea because there was no scientific explanation at that time as to how a supernatural entity could affect a material body.

2. In the 1900s, scientists discovered that the underlying nature of matter is not solid. Rather it is like bundles of the energy of fields. Matter was found to be equivalent to energy. It was discovered that electrons around atoms have various energy states. It was discovered that electrons around atoms can absorb the energy of photons of light which increases the energy state of the electrons.

3. The human brain is composed of billions of neurons, tiny cells that are connected via synaptic gaps and tentacle-like filaments called dendrites.

4. Electrochemical signals are transmitted from neuron to neuron as the synaptic gap "fires" by the transmission of neurotransmitters in synaptic fluid across the synapse.

5. Many scientists consider the on/off firing of brain synapses and the on/off firing of brain neurons to be comparable to the on/off switching of logic gates in a digital computer. Other scientists see the firing of synapses and brain neurons as exhibiting the characteristics of analog computing.

6. Sir John Eccles and Professor Beck identified a quantum mechanical explanation for the firing of the synaptic gaps based on a change in the energy state of the molecules in the gap.

7. One scientifically plausible model of a soul-brain interaction would be for the soul to create miniscule amounts of energy that would change the energy state of the molecules in the synaptic gap, causing the gap to have a higher probability of "firing." The coordinated firing of the appropriate synapses might be able to implement a willful decision by the soul.

8. Another scientifically plausible model would be for the soul to transmit electromagnetic "radio" waves that affect the flow of ions in the dendrites of brain neurons, which would in turn affect the "firing" of the brain neurons.

9. Other mechanisms (such as an unknown noncomputational quantum mechanical action inside microtubules in the cytoskeletal structure of the neurons, quantum coherence, probability fields, and microwaves) might be the manifestation of free will. However, they are not the source of free will because they must act according to rules, even if they are strange, noncomputational rules.

10. Decades of experimental evidence indicates that no mechanism which relies only on matter or energy can be the source of free will because matter and energy must act according to rules. No mechanism which relies only on matter and energy or a nonmaterial entity that is subject to the laws of physics can have a mind of its own.

11. There has been experimental evidence that test subjects report making a decision after a readiness potential in the brain has been observed. This evidence does not imply that the brain has initiated the decision. There are interpretations of the evidence that suggest the free will decisions of the test subjects make use of

the readiness potential to either go forward with or defer the planned action.

12. The ability of people with OCD to use their will to change their behavior is evidence of an ability to go against the quantum probabilities of the brain and to change brain patterns.

"Mathematics is the queen of the sciences."

—*Carl Friedrich Gauss 1777–1855*

"Mathematics, rightly viewed, possesses not only truth, but supreme beauty . . ."

—*Bertrand Russell (1902)* The Study of Mathematics

"The harmony of the world is made manifest in Form and Number, and the heart and soul and all the poetry of Natural Philosophy are embodied in the concept of mathematical beauty."

—*Sir D'Arcy Wentworth Thompson (1917)* On Growth and Form Epilogue

Chapter Seven
MATH

With such a chapter title as "Math," I want to assure the reader that this chapter is not about solving mathematical problems or equations. Rather, we will investigate the meaning of certain mathematical discoveries that occurred during the 1900s and discuss how they provide additional evidence for the existence of supernatural souls.

In his 1994 book, *Shadows of the Mind: A Search for the Missing Science of Consciousness* (*"Shadows"*), Professor Roger Penrose[1] presents an extremely thorough and comprehensive scientific and mathematical investigation of human awareness and thought. To make a valid case that human thought has a supernatural basis, it is necessary to address the issues raised by Professor Penrose in *Shadows*.

In *Shadows*, Professor Penrose describes the seemingly simple and yet far-reaching mathematical theorem that says there are mathematical problems which require human insight and intuition rather than computational solutions. This means that there are mathematical problems that cannot be solved by any set of

---

1. Professor Penrose is the Rouse Ball professor of mathematics at the University of Oxford. He is the author of *The Emperor's New Mind*, which was a New York Times bestseller and was awarded the United Kingdom's 1990 COPUS Prize for science writing. In 1988, he received the internationally prestigious Wolf Prize for physics, shared with Stephen Hawking, for their joint contribution to our understanding of the universe.

mathematical computations and rules. Mathematical computations and rules include operations such as adding, subtracting, multiplying, and dividing as well as logical operations such as "greater than," "less than," and "equal to." Computers use computer programs, which are prescribed sets of mathematical computations and rules,[2] to solve problems. Computers, however, cannot solve problems that require intuition and insight. (Professor Penrose refers to intuition and insight as "noncomputational thinking.") Humans, on the other hand, using intuition and insight can solve problems that require noncomputational solutions, some of which are surprisingly simple. (You always knew you were smarter than computers, didn't you?) Since humans can solve these problems that require noncomputational solutions, human thinking cannot be due solely to a computer-like brain composed of matter that follows prescribed rules.

From the above discussion, I conclude that human thinking (at least the part of human thinking that can solve problems which have noncomputational solutions) must be performed by a non-material (supernatural) entity. However, as described below, Professor Penrose is not yet ready to conclude that this noncomputational thinking requires a supernatural entity. The nature of these problems that require noncomputational solutions and the resulting implications concerning human thought and supernatural human souls are discussed in more detail below.

Human Intuition and Insight

Let us consider a couple of examples of human intuition and insight described by Professor Penrose. One example is to use intuition and insight to prove[3] that:

---

2. Prescribed sets of mathematical computations and rules are also called algorithms.

3. In mathematics it is possible to "prove" things. However, it is first necessary to establish some basic assumptions (called axioms) and then use logical reasoning to develop proofs.

3 x 5 = 5 x 3 and more generally:

A x B = B x A, where "A" and "B" are positive, real numbers.

Professor Penrose suggests we consider an array of dots that has three rows and five dots per row as follows.

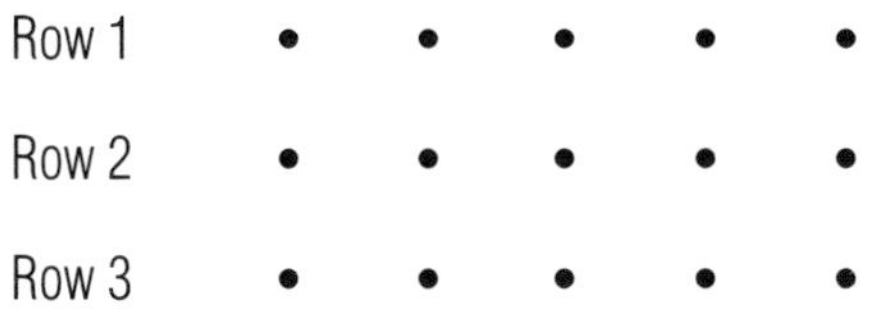

Row 1   •   •   •   •   •

Row 2   •   •   •   •   •

Row 3   •   •   •   •   •

Figure 7.1

Three rows with five dots per row

This configuration of the array represents the calculation 3 x 5. He points out that you could also consider the same array to have five columns with three dots per column.

Columns

1   2   3   4   5

•   •   •   •   •

•   •   •   •   •

•   •   •   •   •

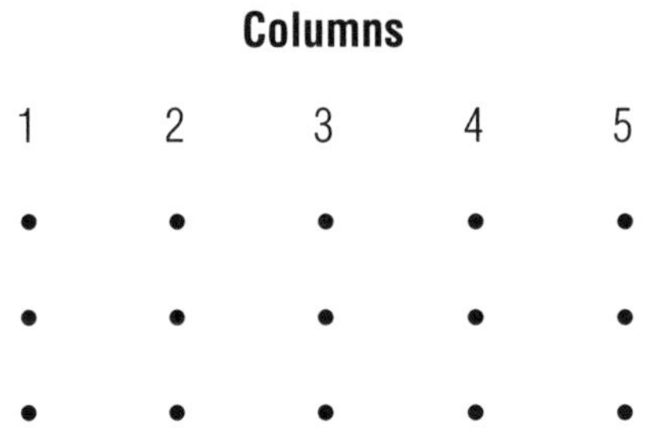

Figure 7.2

Five columns with three dots per column.

This configuration of the array represents the calculation 5 x 3. Since both calculations refer to the same array, this demonstrates, using intuition and insight, that 3 x 5 = 5 x 3. Likewise, you can imagine that the array can have any size. The array could be "A" number of rows with "B" number of dots per row. This configuration of the array would represent the calculation

A x B. You could again consider the same array to have "B" number of columns with "A" number of dots per column. This configuration of the array would represent the calculation B x A. Again, since both calculations refer to the same array, this demonstrates that A x B = B x A.

Both the specific solution (3 x 5 = 5 x 3) and the general solution (A x B = B x A) described above require the mathematician to "see" that the same array can be considered to be either rows or columns. Since we do not know what the actual values of "A" and "B" are, it is not possible to make a numerical calculation for the general case. Rather, the solution is based on insight and not calculation.

Professor Penrose provides a second similar example in which he uses the visual image of looking diagonally at a physical cube from one of its corner points to prove that the sum of hexagonal numbers is always a number that is a cube.[4] Again, he uses intuition and insight, rather than calculation, to validly provide a mathematical proof.

For those who are not able to easily follow some of the algebraic proofs presented by Professor Penrose in *Shadows*, the above two examples provide more tangible examples of what is meant by mathematical intuition and insight. For me, the above examples provide convincing evidence that there are aspects of human thinking that cannot be described in terms of a computer-like brain that makes computations according to prescribed rules. It also provides convincing evidence of the existence of the supernatural, nonmaterial aspects of human thought. Although it

---

4. A hexagonal number is equal to the number of dots in an array of six-sided figures in which each side of the figure is made up of dots. The array is obtained by starting with a single dot in the middle and then surrounding the center dot with six-sided figures, with each figure having one more dot in each side of the figure.

A cube is a number obtained by multiplying the same number times itself three times. For example, 27 is a cube because 27 = 3 x 3 x 3. For a detailed description of this proof, see Penrose (1994), pp. 68–72.

might be difficult to prove mathematically, I do not believe that a computer could have the intuition to discover the relationships described in the above examples or the insight to "see" them after they have been discovered.

NONCOMPUTATIONAL SOLUTIONS

Many mathematical problems can be solved by applying mathematical computations according to established mathematical rules. For example, if you want to know the answer to 2 + 3, you could use a computer and the computer would use a set of established mathematical rules to compute the answer: 2 + 3 = 5. In this case, the computer would use the rules of addition. In general, computers use established mathematical rules to solve problems that have been formulated in mathematical terms. Typically,, mathematical computations and rules are embodied in computer programs (also known as computer software)[5] to manipulate data that is input into the computer.

For some problems, however, there is not any set of mathematical, computational rules that can be set up beforehand that will solve the problems. For these problems, a noncomputational solution is needed, and thus a computer cannot solve the problem.

As Professor Penrose explains in *Shadows*:

> *What I have in mind rests on certain types of mathematically precise activity that can be **proved** to be beyond computation.*[6] *[Emphasis in the original.]*

Let us examine the nature of the problems that require noncomputational solutions and then consider how a mathematician might solve them. Three examples of such problems are listed

---

5. The material that makes up the physical components of a computer such as electronic components, the system board, wires, hard drives, and memory storage devices are know as computer "hardware."

6. Penrose (1994), pg. 28.

below with a short description of each. A more detailed description of the first two examples then follows. The third example is more complex and will not be described in more detail. It is not important that you understand the mathematical structure described in the third example. The main thing to remember is that these are problems for which it has been determined there are no computational solutions. The three examples are as follows:

- Halting problems. There are certain mathematical problems (as described in more detail below) whose solution is to determine if a mathematical computation following logical operations will ever come to a "halt" or will continue forever. In *Shadows*, Professor Penrose presents a proof[7] that halting problems require intuition and insight rather than a computational solution.

- The tiling problem. The tiling problem is to determine whether a group of multisided shapes can completely cover an infinite flat plane surface without gaps and without overlapping the multisided shapes. This was shown to not be solvable by a computational method.[8]

- Hilbert's tenth problem. The problem devised by David Hilbert[9] was to find a computational procedure for determining when both the coefficients and all the solutions of a polynomial equation are integers. In 1970, Yuri Matiyasevich showed there could not be a computational solution to Hilbert's tenth problem.[10] (Don't worry. This is the last you will read about Hilbert's tenth problem in this book.)

---

7. Penrose (1994), pp. 73–75. As described below, there is a different mathematical proof concerning the generalized halting problem.

8. Penrose (1994), pg. 29.

9. David Hilbert was a mathematician who lived from 1862 to 1943.

10. Penrose (1994), pg. 29.

These three problems can have solutions, but the solutions cannot be arrived at by a computational procedure. This means that a human being can figure out solutions to these problems using intuition and insight, but a computer—which operates by following computational rules—cannot. Two of the above problems that require noncomputational solutions are discussed below. They include halting problems and the tiling problem.

Halting Problems

The solution to a "halting" problem involves determining whether or not there is a number or group of numbers that can satisfy a certain type of mathematical question. Halting problems will be described below by an example. If the number or groups of numbers is found that satisfies the mathematical question, then the search can "halt." If it can be determined that there is no number or groups of numbers that satisfy the mathematical question, then it is known that the search would never "halt." The solution to a halting problem is thus either "Yes, there is a number or group of numbers that can satisfy the mathematical question" or "No, there is no number or group of numbers that can satisfy the mathematical question." Note that the solution to a halting problem is *not* the number or groups of numbers that satisfy the mathematical question. The solution to a halting problem, rather, is to be able to answer "Yes" or "No" as described above. For example, consider the following mathematical problem presented in *Shadows*:

Find an odd number that is the sum of two even numbers. [11]

Now I think it is easy to "see" that it is impossible to have an odd number that is the sum of two even numbers. Whenever you add two even numbers you always get an even number. Thus, through human intuition, we can determine that the search to find an odd number that is the sum of two even numbers will

---

11. Penrose (1994), pg. 67.

never stop, that is, it will never "halt." Thus, the answer to the above problem is "No, the search will never halt."

However, if you give this problem to a computer, the only thing the computer can do is to start computing sums to try and find an example of an odd number that is the sum of two even numbers. The computer would begin the following computations of the possible sums that make up odd numbers:

$1 = 1 + 0$

$3 = 1 + 2$

$5 = 1 + 4 \text{ or } 2 + 3$

$7 = 1 + 6 \text{ or } 2 + 5 \text{ or } 3 + 4$

$9 = 1 + 8 \text{ or } 2 + 7 \text{ or } 3 + 6 \text{ or } 4 + 5$

etc.

It is obvious to a human looking at the above pattern that every odd number is the sum of an even number and an odd number. The computer could add numbers to calculate trillions of odd numbers and it will not find one that is the sum of two even numbers. But the computer would not know whether or not the next odd number might be the sum of two even numbers. There is no one computation or set of computations that will show that the search for an odd number that is the sum of two even numbers will not halt. Thus the computer would not be able to conclude that there is no odd number that is the sum of two even numbers. The computer would not be able to intuitively determine that the computations to try and find such an odd number will never "halt." The computer would thus not be able to determine the answer to this halting problem.

Note that to test whether or not an odd number is the sum of two even numbers requires computations. But to determine whether or not the search to find such an odd number will halt requires a noncomputational solution. The answer to a halting problem requires determining whether or not the search for a solution to the mathematical question will end or will continue forever. To solve a halting problem requires human intuition and insight, not a computational procedure.

Another example of a halting problem presented in *Shadows* is the following:

Find a number that is not the sum of four square[12] numbers.

The answer to this problem is not as obvious as the first problem above concerning odd and even numbers. Let us try a few numbers (the same way a computer would try).

$0 = (0 \times 0) + (0 \times 0) + (0 \times 0) + (0 \times 0)$

$1 = (1 \times 1) + (0 \times 0) + (0 \times 0) + (0 \times 0)$

$2 = (1 \times 1) + (1 \times 1) + (0 \times 0) + (0 \times 0)$

$3 = (1 \times 1) + (1 \times 1) + (1 \times 1) + (0 \times 0)$

If you try more numbers you will find that every number can be calculated as the sum of four square numbers. Professor Penrose reports that it can be shown by a mathematical proof, first developed by Joseph Lagrange in 1770, that every number is in fact the sum of four square numbers. However, a computer would just keep calculating away, looking for a number that is not the sum of four square numbers. The computer would not be able to prove that such a number does not exist and that the calculations used to search for it would never halt.

In *Shadows*, Professor Penrose presents a mathematical proof[13] that, in general, halting problems cannot be solved computationally. Note that there is also a general theorem that states there is no computational procedure that can be established to determine whether or not *any* mathematical problem will halt.

---

12. A square number is one that is the product of the same number multiplied by itself. For example, 4 is a square number because $4 = 2 \times 2$.

13. Penrose (1994), pp. 73–75. Professor Penrose recognizes that his proof and subsequent conclusions are controversial and addresses all of the objections of which he is aware.

The Tiling Problem

Anther type of problem that requires a noncomputational solution is the "tiling" problem. For the "tiling" problem, it is necessary to determine if a given shape or shapes would completely cover an infinite flat surface without any overlapping and without leaving any open spaces. There are a number of examples presented in *Shadows*, some of which can tile the infinite plane without leaving spaces and some which do leave uncovered spaces. Consider the following simple example of this type of problem:

If you are given an unlimited number of cross-shaped pieces of tile as shown below, could you completely cover an infinite flat surface without leaving any open spaces?

The start of a solution to the above cross-shaped problem is given below.

Figure 7.3

One cross-shaped piece.

Figure 7.4

Multiple cross-shaped pieces.

I trust that it is fairly obvious that you could keep adding cross-shaped pieces of tile to the above diagram and could completely cover a flat plane without leaving spaces no matter how big it gets. When I say it is "obvious," I am again relying on human intuition and insight that allows us to "see" that making the area larger will not ever result in a different arrangement in which there would be a space left open or in which parts of the tiles would overlap. However, regardless as to how "obvious" the solution to this problem might be, it is not possible to set up a computer with a computer program, no matter how complex, that could determine that the cross-shaped piece can tile an infinite plane. According to Professor Penrose, in 1966 Robert Berger proved that the tiling problem cannot be solved by computational methods.[14]

What Is the Relevance to Souls?

The above examples demonstrate that there are certain problems that require noncomputational solutions using human intuition and insight. The basic question we must now address is: What is the source of human insight to solve the problems that have noncomputational solutions? If humans have brains that are nothing more than biological computers and computers cannot solve noncomputational problems, does this mean our minds must be made up of more than just our material brains? In my opinion, the evidence for noncomputational thinking (intuition and insight) leads directly to the conclusion that human minds must have a nonmaterial, supernatural component that does not have to operate according to prescribed rules. As described by Professor Penrose in *Shadows* and as discussed below, something operating according to prescribed rules would not have the freedom to have the intuition and insight needed to solve problems that have noncomputational solutions. This understanding results from the mathematical theorem of Kurt Gödel which treats noncomputational solutions in a more formal and general way.

---

14. Penrose (1994), pg. 29.

Gödel's Theorem

In *Shadows,* Professor Penrose describes the fundamental mathematical theorem developed by the mathematician Kurt Gödel. In 1931, Mr. Gödel astounded the mathematical world by proposing a new fundamental theorem of mathematics, called the "incompleteness theorem." This theorem states that for any logical system which is powerful enough to generate ordinary arithmetic, there are always logical statements (theorems) that cannot be proved by the axioms of that logical system. Since such a logical system contains theorems that cannot be proved by the axioms of that logical system, that logical system is incomplete, which gives the "incompleteness theorem" its name. Using intuition and insight, however, humans can see the truth of those theorems that cannot be proved by the logical system. In *Shadows,* Professor Penrose paraphrased this new theory as follows:

> *Among the things that Gödel indisputably established was that no **formal system** of sound mathematical rules of proof can ever suffice, even in principle, to establish all the true propositions of ordinary arithmetic. This is certainly remarkable enough. But a powerful case can also be made that his results showed something more than this, and established that human understanding and insight cannot be reduced to any set of computational rules. For what he appears to have shown is that no such system of rules can ever be sufficient to prove even those propositions of arithmetic whose truth is accessible, in principle, to human intuition and insight—whence human intuition and insight cannot be reduced to any set of rules. It will be part of my purpose here to try to convince the reader that Gödel's theorem indeed shows this, and provides the foundation of my argument that there must be more to human thinking than can ever be achieved by a computer, in the sense that we understand the term "computer" today.*[15] *[Emphasis in the original.]*

---

15. Penrose (1994), pp. 64–65.

What Professor Penrose is saying is that Gödel's theorem implies there are aspects of human thinking and thought that cannot be explained in terms of how a computer operates. Human intuition and insight can determine the truth about propositions of arithmetic that cannot be proved by a set of rules. Thus, "human intuition and insight cannot be reduced to any set of rules." For a more detailed discussion of the implications of Gödel's theorem, see the Appendix to this chapter.[16]

Because a computer acts according to physical principles and computational rules embodied in its hardware and software,[17] a computer does not have the ability to have the human intuition and insight which transcends mechanical computation. Professor Penrose explains further in Chapter 1 of *Shadows*:

> *Central to the arguments of Part I [of* Shadows*], is the famous theorem of Gödel, and a very thorough examination of the relevant implications of Gödel's theorem is provided. . . . The conclusions are that conscious thinking must indeed involve ingredients that cannot be even **simulated** adequately by mere computation; still less could computation, of itself alone, evoke any conscious feelings or intentions. Accordingly, the mind must indeed be something that cannot be described in any kind of computational terms.[18] [Emphasis in the original.]*

In this excerpt, Professor Penrose is explaining that computations alone cannot adequately simulate conscious human thought.

---

16. A discussion of Gödel's theorem can also be found in Stephen Barr's 2003 book *Modern Physics and Ancient Faith*.

17. Computer hardware is made up of the solid parts of a computer such as the circuit boards and memory storage devices. The software is made up of the programs stored in memory that direct the computational activities of the computer. At its most basic level, however, software is also hardware because software is the arrangement of the hardware memory in digital states of "1's" or "0's". Each program has its own arrangement of "1's" and "0's" in the hardware memory.

18. Penrose (1994), pg. vi of preface.

When Professor Penrose uses the term "computation," he is referring to the operation of a computing device (a computer) which follows rules of computation. These rules of computation can be as complex as you would like to make them, yet they would still not be able to simulate conscious human thinking.

Thus, it seems logical to me to conclude that the human brain, which is also subject to physical laws and operates as a very complex biological computer, is not sufficient to explain human intuition and insight. For me, the concept that "human intuition and insight cannot be reduced to any set of rules" as stated in the above quotation, leads directly to the conclusion that we have to look beyond the biological computer known as the human brain. However, Professor Penrose is not ready to accept this conclusion.

Rather, he is intent on finding a "scientific" basis for understanding how the human mind works rather than jumping to what he considers a "mystical" explanation. He supports instead "a genuine search for a means, within the constraints of the hard facts of science, whereby a scientifically describable brain might be able to make use of subtle and largely unknown physical principles in order to perform the needed noncomputational actions."[19] Professor Penrose understands very well the importance of being able to explain human consciousness in scientific terms. As he states in *Shadows*:

> *A scientific worldview which does not profoundly come to terms with the problem of conscious minds can have no serious pretensions of completeness.*[20]

As described above, his approach is to look for "unknown physical principles" operating in the brain to explain human thought. He recognizes that our present understanding of physics

---

19. Penrose (1994), pg. v of preface.

20. Penrose (1994), pg. 8. It is important to remember this comment when reading about the Theory of Everything in the chapter "Quantum Mechanics."

does not provide a mechanism that can explain even conceptually how matter can function in such a way as to produce the noncomputational aspects of human thought that provide intuition and insight. Thus he prefers to believe that science will someday be able to explain noncomputational thought as a natural phenomenon rather than to conclude that science is not only *currently* unable to explain noncomputational thought in physical terms but that science will *never* be able to explain noncomputational thought in physical terms.

Based upon the amazing scientific discoveries in the 1800s and 1900s, Professor Penrose's approach seems reasonable. The scientists of the 1700s did not know about the invisible electromagnetic waves which form the basis of radio, television, and microwave communication which are commonplace today. They likewise did not understand the composition of an atom and how it might be split to release enormous quantities of energy or be fused with other atoms to also release energy in the same type of reaction that powers the sun. Should we not give scientists a chance to discover some "unknown physical principle" which is the basis for human thought? As described below, as long as the physical matter of the brain follows scientific laws, even it they are noncomputational laws, it will never be able to explain noncomputational thinking that provides human insight.

Professor Penrose seeks a "scientific" answer because he believes that science and math have been the source of all the major advancements in understanding how the physical world works. He defines four potential explanations for understanding human consciousness and thought, one of which, "D," is a nonscientific or "mystical" explanation. He rejects the "mystical" explanation "D" because it is not scientific. As he states in *Shadows*:

> *My own reasons for rejecting D—the viewpoint which asserts the incompetence of the power of science when it comes to matters of the mind—spring from an appreciation of the fact that it has only been through use of the methods of science and mathematics that any real progress in understanding the*

behaviour of the world has been achieved. Moreover, the only minds of which we have direct knowledge are those intimately associated with particular physical objects—brains—and differences in states of mind seem to be clearly associated with differences in the physical states of brains. If it were not for the puzzling aspects of consciousness that relate to the presence of "awareness" and perhaps of our feelings of "free will," which as yet seem to elude physical description, we should not need to feel tempted to look beyond the standard methods of science for explanation of minds as a feature of the physical behaviour of brains.[21]

Unfortunately, Professor Penrose has engaged in a bit of circular reasoning. He rejects "D" a supernatural or "mystical" explanation and favors a scientific or mathematical explanation because it has been through science and math that we have made progress in understanding the physical world. Science and math would therefore be useful in explaining human thought as a natural phenomenon only if human thought is in fact a completely physical phenomenon. But that is the very thing that is under investigation, that is, whether we can explain human thought in purely physical terms or whether we need a nonphysical, supernatural explanation. Thus, the rejection of "D" is, at best, premature.

Professor Penrose's final concluding "If" statement is particularly revealing. He admits that "awareness" and "free will" cannot be shown to have a physical origin. This in itself indicates a supernatural explanation should not yet be rejected. By putting the "If" statement at the beginning of the paragraph, we could paraphrase his reasoning as follows:

If human awareness and free will are not supernatural phenomenon, then they are physical phenomenon and can be explained by science and math. Science and math are the basis for our understanding of physical phenomenon. Therefore,

---

21. Penrose (1994), pg. 50

(even though human awareness and free will cannot yet be explained as a physical phenomenon), I reject a supernatural explanation.

As described above, it appears he has assumed the premise that we are trying to prove. He has assumed that human awareness and free will are not supernatural phenomenon. It is important to remember that science is concerned with understanding the rules of how the material and natural universe operates. Science does not claim to be able to explain how supernatural reality operates. Science does not claim there is no supernatural reality (although some scientists might seem to imply that it does). Thus, if there is a supernatural aspect of human thought, science will not be able to explain it. We should not reject a supernatural explanation just because science is not able to explain it. The fact that science cannot explain supernatural reality (and specifically claims that it can *only* explain the material and natural universe) does not prove that there is no supernatural reality.

Conversely, believing in a supernatural reality does not negate all of the benefits that science and scientific thinking has brought to humankind. Both supernatural reality and material reality are completely compatible. Also, it is important to remember that science is just a tool. It is by free will that we decide how to use that tool. Science in and of itself does not provide the material wealth that we enjoy. I would dispute the contention that "it has only been through use of the methods of science and mathematics that any real progress in understanding the behaviour of the world has been achieved." This statement is only true if we do not include the behavior of human beings as being a part of the "behaviour of the world" that is now understood. Certainly the methods of science and mathematics have allowed us to understand how the material world "behaves," and we have used this understanding to develop incredible material wealth. However, I would submit that understanding spiritual and moral values (or lack thereof) provides a better explanation and understanding of the behavior of human beings in this world. Likewise, it is spiritual and moral values that have provided a foundation for allowing science to be used for good rather than for evil.

Understanding scientific laws does not help us understand how humans will use knowledge of the material world. Scientific discoveries can be used for good or evil, for peace or war, for justice or injustice, and to free or to enslave people. Only by knowing the spiritual and moral character of the people that control scientific knowledge will we be able to understand how it might be used. As long as people have free wills, they will be able to use any institution or knowledge for good or for evil. Science can be used to kill, injure, and leave homeless millions of people as has been done in countless wars, or it can be used to improve the lives of people.

I agree with Professor Penrose that various states of mind seem to be related to physical states of the brain. For example, drugs and alcohol, as well as physical trauma (such as a blow to the head), can affect the brain and the states of the mind. This, however, is not a sufficient reason to conclude that all mental processes have a physical basis. In fact, his very next sentence (the "If" statement") admits that human awareness and free will cannot yet be explained as a physical phenomenon. What puzzles me is the rejection of the possibility of there being a supernatural explanation even after such an admission. The logical approach would be to keep the supernatural possibility open until it can be shown that a supernatural explanation *does not work* and that human thought *can* be completely explained as a physical phenomenon.

SCIENTIFIC RULES CANNOT EXPLAIN NONCOMPUTATIONAL THOUGHT THAT PROVIDES INTUITION AND INSIGHT

Having experiments that are reproducible is the core aspect of the scientific method. This is the very quality that Professor Penrose finds so appealing in science and the reason science has been so successful in describing how the material world operates. It is because of this predictability that a scientist can mix two chemicals together or an engineer can design a bridge, a computer, a car, or a telephone system and be confident of the results.

The noncomputational thought that provides intuition and insight requires something other than the reproducibility required by the scientific method. If human thought is only a product of the brain which is made up of matter which moves and reacts according to the laws of physics, then noncomputational thought that provides intuition and insight would not be possible. A human brain, acting according to the laws of physics, is like a computer that operates according to certain prescribed rules. A computer is only capable of following the prescribed rules and is thus only capable of computational thought. I think I am in agreement with Professor Penrose on this matter as he says in the excerpt earlier in this chapter that "human intuition and insight cannot be reduced to any set of rules."[22]

For there to be a scientific explanation of human thought (including noncomputational thought that provides intuition and insight) which relies only on the material world, there would need to be some type of mechanism for particles to choose what they will do and "have minds of their own."

Professor Penrose recognizes that quantum mechanics cannot explain the noncomputational nature of consciousness. In *Shadows*, he states:

> *Thus, it appears that neither classical nor quantum physics, as presently understood, allows room for noncomputable behaviour of the type required [to explain human intuition and insight], so we must look elsewhere for our needed noncomputable action.*[23]

---

22. See the first excerpt after the heading "Gödel's Theorem."
23. Penrose (1994), pg. 216.

It is also interesting to note that Professor Penrose considers the underlying mathematical formulas that describe quantum mechanics to be as deterministic as other scientific theories.

> *The descriptions [of the quantum-level world] are per-fectly clear cut—and they provide us with a micro-world that evolves according to a description that is indeed math-ematically precise and, moreover,* **completely deterministic!**[24] *[Emphasis in the original.]*

And also:

> *... the mathematical laws that govern the quantum world are remarkably precise—as precise as the more familiar equa-tions that control the behaviour of macroscopic objects—despite the fuzzy images that are conjured up by such descriptions as "quantum fluctuations" and "uncertainty principle."*[25]

As Professor Penrose explains, the underlying theory and mathematical description of quantum mechanics is precise and deterministic. However, due to the uncertainty principle, it is not possible to measure atomic and subatomic particles without affecting their position and motion. The probabilities normally associated with quantum mechanics occur when measurements of particles are made. As described by Professor Penrose:

> *... it is only ... when measurements are made, that inde-terminacy and probabilities come in. A measurement of a quantum state occurs, in effect, when there is a large magnifi-cation of some physical process, raising it from the quantum to the classical level.*[26]

As described above, neither a deterministic model of the uni-verse (as described by classical physics and the underlying for-mulas of quantum physics) nor a probabilistic model (as described

---

24. Penrose (1994), pg. 259.
25. Penrose (1994), pg. 313.
26. Penrose (1994), pg. 264.

by quantum physics when measurements are made) can even conceptually explain noncomputational thinking that provides for intuition and insight.

If these models of the universe are replaced or augmented by a new scientific model, it will only be because that new model has been subjected to the scientific method and is able to consistently produce the same results given the same initial conditions. For a model of the universe to be scientific and produce consistent results, it must be made up of a consistent set of rules. Any model that consists of matter or energy that follows rules will not be able to explain noncomputational thought that provides intuition and insight. It is my belief that intuition and insight must be caused by something that does not have to follow the laws of physics.

Professor Penrose, on the other hand, has a fairly specific idea as to the nature of the physics which he believes is used by the brain to produce the noncomputational thoughts described earlier.[27] He believes the required noncomputational theory will be found in a theory of physics that can explain both the underlying deterministic behavior of subatomic particles predicted by the quantum wave equations as well as the probabilistic behavior apparent in the measurement of subatomic particles as described previously.

Professor Penrose names the unknown noncomputational physics theory that might be able to explain human consciousness as the objective reduction (OR) theory. The OR theory has not yet been defined, but it will be a theory that replaces the reduction procedure (the "R-procedure" which is Professor Penrose's terminology for the collapse of the wave function) which currently describes the probabilistic behavior of subatomic particles.[28] He considers the R-procedure to be only a temporary or

27. See the chapter "The Soul-Brain Interface" for a description of where Professor Penrose theorizes the OR activity might be located in the brain.

28. See the chapter "Quantum Mechanics" for a description of the collapse of the wave function resulting in the probabilistic nature of matter.

partial theory which is acting as a "stop-gap" theory until a more complete understanding is provided by the OR theory. The OR theory will be a theory that resolves the inconsistency between the deterministic and probabilistic behavior of subatomic particles. He describes this theory of physics which will be able to explain noncomputational thinking as follows:

> *This physics is the missing OR theory that straddles the quantum and classical levels and, as I am arguing, replaces the stop-gap R-procedure by a highly subtle noncomputational (but undoubtedly still mathematical) physical scheme.*[29]

Since the OR theory has not yet been defined, it is not possible to comment directly on it. It is clear Professor Penrose expects the OR theory of physics to be a true scientific theory and to be based on something that exists in the physical world. He wants it to be able to follow mathematical rules and be deterministic even though the rules would be noncomputational.

I doubt that matter follows noncomputational rules. However, we could hypothesize that scientists someday find a physical process that follows noncomputational rules as proposed by Professor Penrose. Such noncomputational rules, however, would not be able to explain noncomputational thinking. The solution to a problem that requires a noncomputational solution requires human insight. However, something that is following noncomputational rules is just following rules. It is not providing insight. As Professor Penrose says himself: "human intuition and insight cannot be reduced to any set of rules."

Dual Nature

With human free will being explained by a supernatural soul, the complete human being has a dual nature: a physical nature associated with the body and a supernatural nature associated with the soul. Professor Penrose, however, believes there are some dif-

---

29. Penrose (1994), pg. 373.

ficulties with the dualistic body-mind concept for human beings. As he explains in *Shadows*:

> In my own opinion, it is not very helpful, from the scientific point of view, to think of a dualistic "mind" that is (logically) **external** to the body, somehow influencing the choices that seem to arise in the action of R. If the "will" could somehow influence Nature's choice of alternative that occurs with R, then why is an experimenter not able, by the action of "will power," to influence the result of a quantum experiment? If this were possible, then violations of the quantum probabilities would surely be rife! For myself, I cannot believe that such a picture can be close to the truth. To have an external "mind-stuff" that is not itself subject to physical laws is taking us outside anything that could be reasonably called a scientific explanation, and is resorting to the viewpoint D.[30] [Emphasis in the original.]

Responding to his last comment first, he appears to be rejecting the existence of "external mind-stuff" simply because it is not subject to physical laws and reflects viewpoint "D," which is the "mystical" explanation. Once again, he is simply assuming that the only things that exist are those that follow scientific laws. It is true that science does not claim to be able to explain supernatural reality. But it is not logical to discard the possibility of a supernatural explanation just because science cannot explain how a supernatural entity functions. It is the very nature of free will that it cannot be subject to scientific rules. If it were subject to rules, free will would not be free.

Ironically, not believing in supernatural souls requires a more "mystical" explanation in order to explain free will. This is because without supernatural souls, for free will to be possible, the matter that makes up the human brain would have to be free to not interact in accordance with the laws of physics.

---

30. Penrose (1994), pg. 350.

The fact that the mind is not subject to physical laws is the reason an external, supernatural mind is able to explain the phenomenon of human free will. It is thus a benefit that the mind is not subject to physical laws. I am in complete agreement that this is the viewpoint "D," which relies on a supernatural[31] explanation of human consciousness and free will. I do not agree, however, that the existence of a supernatural soul implies that an experimenter should be able to influence the result of a quantum experiment using "will power." The soul can very logically be confined to interacting with just one human brain. A potential basis for this limitation is discussed in the chapter "The Soul-Brain Interface."

With the soul limited to interacting only with the matter of one human brain, it would not be able to directly influence anything outside the body. The ability of humans to influence their surroundings is limited to interactions based on using the various bodily senses. The existence of human souls, thus, does not imply that violations of quantum probabilities would be "rife."

Note also, as described in the chapter "Free Will or Not," that Professor Penrose might consider human free will to be an illusion. It is not clear whether or not he actually believes that and, if so, how much that affects his conclusions.

Cognizers

In their 1988 book *Cognizers: Neural Networks and Machines that Think,* R. Colin Johnson and Chappell Brown discuss another aspect of human thinking that they call "noncomputational" thinking. However, they do not use the term "noncomputational" to refer to the thinking required to arrive at the types of noncomputational solutions described at the beginning of this chapter. Rather, they are referring to the mechanism they theorize enables humans to learn. The authors consider the ability to

---

31. I think the word "mystical" might have some negative connotations to some people.

learn to be the main characteristic that separates humans from current computers.

> *The major difference between computers and humans is that computers are programmed, whereas humans learn.*[32]

They describe a new generation of electronic machines called "cognizers" that would have intelligence comparable to humans. Cognizers would surpass the abilities of computers which can only compute. Cognizers would also be able to learn:

> *Though digital computers are being used to simulate the information processing of the brain, machines that truly model the brain, rather than merely simulate it, substitute noncomputational physical mechanisms for the logic and mathematics of computation.*[33]

The authors believe the critical feature of cognizers would be the ability to learn in the same way that humans do:

> *Cognizers perceive semantic content in precisely the same manner that humans do—that is, by learning from experience. What they learn about are the causal connections among objects so that they can generalize in the future with intelligence wrought from the past.*[34]

The authors accept the theory that humans learn by having neural networks strengthened by certain behaviors and responses to the environment. They explain how electronic machines could also "learn" like humans by similarly having electronic circuits that are strengthened by certain behaviors and responses. This method of learning is the "noncomputational" mechanism referred to above.

There are, however, characteristics other than the ability to learn that distinguish humans from computers or cognizers. The

---

32. Johnson and Brown (1988), pg. 42.

33. Johnson and Brown (1988), pg. 5.

34. Johnson and Brown (1988), pg. 15.

authors, for example, do not show how a computer (or a cognizer) can have intuition and insight and perform the noncomputational thinking required to solve the mathematical problems discussed above in this chapter. Neither do they explain how any electronic machine that is only made out of matter can make free choices.

Chapter Conclusions

1. It has been proven that there are mathematical problems that cannot be solved by computational methods.

2. Gödel's theorem (which has been proven) implies that there is human mathematical thought that cannot be reduced to a set of computational rules.

3. Computers can only solve problems using computational methods.

4. Computers cannot solve noncomputational problems.

5. Using insight and intuition, humans can solve problems that require noncomputational solutions.

6. Human insight and intuition can recognize the truth and validity of the Gödel theorem. Computers cannot.

7. Human insight and intuition cannot be simulated by computational rules.

8. Computers rely on material facilities to make their computational solutions.

9. Human thought is conceptually different from the operation of computers.

10. The scientific method requires that experiments be reproducible for identical circumstances.

11. Experimental evidence indicates that all matter and energy act and move according to scientific laws and follow established rules.

12. Experimental evidence indicates that matter and energy cannot act and move with "minds of their own."

13. The mental processes that solve noncomputational problems cannot be explained using anything that is subject to scientific laws, even if we do not know what those laws are and even if those laws are noncomputational.

14. The human brain, which is made up of matter, operates according to scientific laws and follows rules, even if we do not know what those laws and rules are. The brain cannot on its own explain human free will and the insight required to solve noncomputational problems.

15. The discovery of a noncomputational physical process would not provide an explanation for the solution to noncomputational problems. The solution to noncomputational problems requires insight. Noncomputational problems cannot be solved by following rules, even if they are noncomputational rules.

16. There must be some force that does not originate with the material body or brain which affects the matter that makes up the brain and is the source of insight to solve noncomputational problems.

17. If such outside force is supernatural and is not governed by the rules of science, it can act freely and be the source of insight to solve noncomputational problems.

18. A supernatural source, such as a soul, can be limited to interact with a limited amount of matter such as the brain of a person. A possible mechanism for this limitation is discussed in the chapter "The Soul-Brain Interface."

19. There can be no scientific law (existing or future) that establishes rules for the action and movement of matter and energy that will be able, even conceptually, to provide solutions to noncomputational problems.

20. It would be unscientific to believe that a brain, which is made up of matter that must operate according to scientific laws (even noncomputational laws), could by itself enable a human being to have the insight to solve noncomputational problems.

21. Computers act according to computational rules and cannot be developed to perform noncomputational "thinking" regardless how complex the operating system or computer program is.

Appendix to Math Chapter

Gödel's Incompleteness Theorem

As described in the chapter "Math," Gödel's theorem implies that no formal logical system that is powerful enough to generate ordinary mathematics can establish all the true propositions of arithmetic. There will be some true propositions it will not be able to establish. Thus, such a formal logical system will be incomplete. However, humans using intuition and insight will be able to see the truth in those arithmetic propositions that the formal logical system cannot establish. Also, humans using intuition and insight can see the validity of the Gödel theorem, which a device that uses a formal logical system cannot do. Thus, humans using intuition and insight can do more than something that is limited to using a formal logical system. Since computers are based on using a formal logical system, humans using intuition and insight can do more than computers can.

In his 1997 book *Mind Matters: Exploring the World of Artificial Intelligence*, James Hogan reports that the implication of Gödel's theorem described above was first identified by philosopher John Lucas in a 1963 book *The Modeling of Mind: Computers and Intelligence* "as proof that computing machines would never be able, on principle, to match human thinking."[35]

Mr. Hogan also reports that he asked Professor Marvin Minsky, one of the pioneers in the research of artificial intelligence, to address this limitation on computers being able to match human thinking:

> *When I put this to Marvin Minsky, he suggested that a more illuminating way of viewing Gödel's conclusion might be that any formal system operating under the constraint of being consistent would be too weak to support the familiar forms of commonsense human reasoning, particularly with regard*

---

35. Hogan (1997), pg. 288.

to "reflexive" processes, in which thoughts are directed toward themselves. What this means is that to make machines appear reasonably smart in the ways we humans take pride in being smart, one must use logically defective shortcuts (i.e., "heuristics"). This isn't particularly difficult to program machines to do, but the results wouldn't be acceptable in any of the typical application roles we're accustomed to seeing for computers, wherein the whole point of using them lies in their predictability and repeatability and anything short would be considered tantamount to worthlessness. It would mean that when interpreting what they were doing in commonsense terms, we would find the "logic" they were using inconsistent. This is another way of saying that sometimes they'd get things wrong and other times contradict themselves—just like people.[36]

Assuming that Mr. Hogan has accurately described Professor Minsky's response, I do not think it refutes the conclusion reached by Professor Penrose and John Lucas. Gödel's theorem does not say anything about the fact that people sometimes make logical mistakes. The theorem says that certain logical systems are incomplete, not that they make mistakes. Humans using intuition can see the truth of this theorem and the truth of arithmetic propositions that a logical system cannot establish because it is incomplete. Simply programming a computer to make logical mistakes does not give it the ability to see the truth that humans can see. As Professor Penrose explains:

> *But why, then, can one not simply get a computer also to follow this Gödel argument and itself "see" the truth of any new Gödel proposition? The catch lies in "seeing" that the Gödel argument, in any specific realization, has actually been correctly applied. The trouble is that the computer does not have a way of judging truth; it is only following rules. It does not "see" the validity of the Gödel argument. It does not "see" anything unless it is conscious! It seems to me that in order*

---

36. Hogan (1997), pp. 288–289.

*to appreciate the validity of the Gödel procedure—or, indeed, to see the validity of **any** mathematical procedure—one must be conscious. One can follow rules without being conscious, but how does one **know** that those rules are legitimate rules to follow without being, at least at some stage, conscious of their meaning?[37] [Emphasis in the original.]*

Another way of understanding Professor Penrose's observation is to recognize, as Professor Minsky suggests, that computers can be programmed to make logical mistakes or can be programmed to be completely logical and consistent. As Professor Penrose comments, however, a computer does not *know* or *understand* what is valid logic and what is a mistake in logic. The computer is just following the rules that are programmed into it.

I expect that the response of the artificial intelligence community will be that humans also do not know what is valid logic and what is a mistake in logic unless they are told. Thus, they would say we are "programmed" to accept some statements as being logical and some as being illogical. But I do not think this is correct. First of all, some humans, based on reason, intuition, and insight had to initially determine (i.e., choose) what is logical and what is not logical. A computer could not do this. Then, we each have to choose whether or not to accept that logic as being valid. Note again the element of choice in this process. Our decision as to whether or not to accept a system as being logical is based on our ability to reason, but is also based on human intuition as to what is logical and what is not. However, even if the artificial intelligence community does not accept this explanation of human reasoning and logic, the observation remains valid that computers cannot see the truth of the Gödel theorem, while humans can.

---

37. Blakemore (1987), pg. 270.

"All's Well that Ends Well"

—*William Shakespeare*

Chapter Eight
CONCLUSION

There are various types of conclusions:

- **The end of a story:** They got married and lived happily ever after.

- **A summary:** In conclusion, we can see that the available evidence indicates that motionless, unconstrained, electrically neutral matter will begin to move toward other matter.

- **A logical inference:** From the evidence, I conclude that there is something (which has been named gravity) which causes motionless, unconstrained, electrically neutral matter to move toward other matter.[1]

- **A practical decision:** Since there is gravity, I conclude that I will not jump from high places.

THE END OF THE STORY

In life a wedding is not the end of a story but is the beginning of the really interesting and fulfilling part of life. Similarly,

---

1. As described in Chapter 1, according to the theory of Newton, gravity is a force which causes matter to attract other matter. According to Einstein's theory of relativity, matter warps space and time.

I hope that this book is not the end of the story concerning the hypothesis that human free will cannot be explained as a natural phenomenon that is subject to the laws of physics and chemistry and that free will can be explained by people having supernatural souls. I hope this issue and the evidence described herein will be discussed in many scientific and nonscientific classrooms and other venues throughout the world.

A Summary of Evidence

The main evidence on which I have relied to support the hypothesis that free will cannot be explained as a natural phenomenon and that all humans have supernatural souls is the following:

- The evidence from decades of scientific experiments that indicates all natural phenomena (atoms, molecules, energy, and forces) interact according to laws and
- The evidence that humans have free will.

Additional supporting evidence is that humans can perform noncomputational thinking and that people with obsessive-compulsive disorder can use mental force to change their behavior.

Logical Inferences

From the above evidence, I conclude that if humans have free will, the source of free will cannot be a natural phenomenon that is subject to the laws of nature. For free will to be truly free, the source of free will must be a supernatural entity that is not subject to the laws of nature. If it is assumed humans do not have this supernatural entity, then the logical conclusion is that humans do not have free will. These conclusions are based on logical reasoning and are a part of my belief system.

Most scientists recognize all of the evidence to date indicates all natural phenomena are subject to the laws of nature. Based on this evidence, some scientists conclude that there is no supernatural reality. However, the inability to directly detect something does not mean it does not exist. No one has ever seen an electron wave around the nucleus of an atom but scientists conclude (and believe) it is there based on evidence. No one has ever seen virtual particles, but scientists conclude (and believe) they exist based on the effect they have on other particles. Similarly, in my opinion, the evidence summarized previously provides evidence for a supernatural reality.

Some scientists have concluded that humans do not have free will based on the evidence that all natural phenomena are subject to laws. However, such a conclusion is based on the unstated assumption that anything that cannot be explained as a natural phenomenon does not exist and the assumption that there is no supernatural reality. Science, however, is the study of natural phenomena and does not claim there is no supernatural reality. In addition, there is evidence in the form of a test that allows people to conclude (and believe) humans have free will. Of course, it is absurd to freely choose to not believe in free will because, without free will, it would not even be possible to freely choose whether or not to believe in free will.

If free will does not exist, it is likewise absurd for me to "try to convince" people that they have free will because I could not freely choose what I try to do or say, and no one could freely choose what he or she will believe. Without free will, life is meaningless because humans would not have control over what they do or even what they think. In my opinion, incorrectly using scientific principles to conclude that free will does not exist is truly ironic because the scientific method requires free will to choose hypotheses, to make decisions about the types of experiments to perform, and to make decisions about the validity of evidence.

Science has helped clarify the difference between the natural and the supernatural worlds:

- **The natural world is made up of entities that are subject to laws.**

- **The supernatural world has entities that have free will and are not subject to laws.**

There are several hypotheses based on sound scientific principles which provide theories as to how a supernatural entity can interact with a material brain. There are no scientific theories as to how a natural phenomenon subject to laws can be the source of free will.

Both computers and robots, which operate using computer programs, cannot perform noncomputational thinking. Although some computers and robots might be programmed to "learn" from their experiences, they cannot be programmed to have free will and freely choose what they "want" to do regardless of what the program is instructing them to do.

Practical Decisions

The belief that all humans have supernatural souls leads to many practical conclusions that can also be a part of a person's belief system. Each person must decide based on his or her value system how he or she will incorporate such belief into his or her life. For me, belief in human supernatural souls leads to the following conclusions and expectations.

An almighty, supernatural Being (I will call "God") is needed to create supernatural souls and join them to human bodies in a miraculous way. This implies that a God who loves us put humans on earth for a purpose and that we should decide to respect the life, liberty, and human rights given by God to all people (at all stages of their lives) so that they have the freedom to accomplish the purpose given them by God. A supernatural soul makes each human a child of God and a creature who has God-like powers such as the ability to make free choices, to love, to choose good over evil, to forgive, to create, and to feel joy.

I believe the need for a supernatural entity as the source of human free will makes the controversy about the theory of evolution less critical. The important characteristic about humans is that human souls are made in the image of God. It is irrelevant to me whether human bodies are the result of "special creation" or were created by millions of years of evolution. I would hope that everyone can accept the evidence for free will and supernatural souls while the evidence for evolution, intelligent design, and/or special creation is further studied and refined.

I believe that human spirituality will be enhanced as more people become aware that being aware of being aware is not a natural phenomenon. The first step in enhancing spirituality is recognizing that humans are primarily supernatural (spiritual) beings and then making choices during life based on that recognition.

It is my hope:
- that all people can recognize there is overwhelming evidence that leads to the conclusion that they have supernatural souls;
- that this recognition and the hope for eternal life will help relieve at least in part the depression and suffering experienced by some people;
- that belief in a supernatural reality and a supernatural Being is a source of healing for guilt;
- that people will realize how wonderful free will, life, and existence are; and
- that these realizations will result in an attitude of awe and thankfulness and will renew the joy of living in many people.

Finally, I hope that a recognition that each human soul is made in the image of a spiritual God will help human relations at all levels and lead to a spiritual millennium.

Bibliography

Adler, Mortimer J. 1967, 1993 ed. *The Difference of Man and the Difference It Makes.* New York: Fordham University Press.

Adler, Mortimer J. 1985. *Ten Philosophical Mistakes.* New York, NY: Macmillan Publishing Company.

Adler, Mortimer J. 1990. *Intellect: Mind over Matter.* New York, NY: Macmillan Publishing Company

Aquinas, Thomas. 1273. *Summa Theologiae,* 1952 ed. Daniel J. Sullivan, ed. Chicago: Encyclopedia Britannica, Inc.

Ardrey, Robert. 1961. *African Genesis.* New York: Dell Publishing Co., Inc.

Aristotle. 1986. *De Anima (On the Soul).* London /New York: Penguin Books.

Aristotle. 1947. *Introduction to Aristotle.* Richard McKeon, editor. New York: Random House The Modern Library.

Barbour, Ian G. 2000. *When Science Meets Religion.* SanFrancisco: Harper.

Barr, Stephen M. 2003. *Modern Physics and Ancient Faith.* Notre Dame, Indiana: University of Notre Dame Press.

Behe, Michael. 1996. *Darwin's Black Box: The Biochemical Challenge to Evolution.* New York: The Free Press.

Bettelheim, Bruno. 1982. *Freud and Man's Soul.* New York: Alfred A. Knopf Inc.

Blakemore, Colin and SusanGreenfield, ed. 1987. *Mindwaves: Thoughts on Intelligence, Identity, and Consciousness.* Cambridge: Basil Blackwell.

Boyd, Gregory A. and Edward K. Boyd. 1994. *Letters from a Skeptic A Son Wrestles with His Father's Questions about Christianity.* Wheaton, Illinois: Victor Books.

Brown, Michael H. 1990. *The Search for Eve.* New York: Harper & Row.

Brown, Warren S., Nancy Murphy, and H. Newton Maloney, ed. 1998. *Whatever Happened to the Soul? Scientific and Theological Portraits of Human Nature.* Minneapolis: Fortress Press.

Browne, Janet. 2002. *Charles Darwin The Power of Place*. New York: Alfred A. Knopf.

Chalmers, David J. 1996. *The Conscious Mind: In Search of a Fundamental Theory*. New York: Oxford University Press.

Chopra, Deepak, M.D. 1993. *Ageless Body, Timeless Mind: The Quantum Alternative to Growing Old.* New York: Harmony Books.

Crick, Francis. 1994. *The Astonishing Hypothesis: The Scientific Search for the Soul*. New York: Charles Scribner's Sons.

Csikszentmihalyi, Mihaly. 1997. *Finding Flow The Psychology of Engagement with Everyday Life*. New York, New York: Basic Books HarperCollins Publishers, Inc.

Damasio, Antonio R. 1994. *Descartes' Error: Emotion, Reason, and the Human Brain*. New York: A Grosset/Putnam Book.

Darwin, Charles. 1859. *The Origin of Species by Means of Natural Selection or The Preservation of Favoured Races in the Struggle for Life*. London: John Murray.

Davies, Paul. 1999. *The Fifth Miracle The Search for the Origin and Meaning of Life*. New York: Simon and Schuster.

Davis, P. Williams. 1989. *Of Pandas and People: The Central Questions of Biological Origins*. Dallas, Texas: Haughton Publishing Co.

Dawkins, Richard. 1976, 1989 ed. *The Selfish Gene.* New York: Oxford University Press.

Dawkins, Richard. 1982. *The Extended Phenotype. The Gene as the Unit of Selection.* Oxford: Oxford University Press.

Dawkins, Richard. 1986. *The Blind Watchmaker: Why the Evidence of Evolution Reveals a Universe Without Design*. New York: W. W. Norton & Company.

Dembski, William A. 1998. *Mere Creation: Science, Faith and Intelligent Design.* Downers Grove, IL: Intervarsity Press.

Dembski, William A. 2002. *No Free Lunch: Why Specified Complexity Cannot Be Purchased without Intelligence*. New York: Rowman & Littlefield Publishers, Inc.

Dennett, Daniel C. 1984. *Elbow Room The Varieties of Free Will Worth Wanting.* Cambridge, MA: A Bradford Book The MIT Press.

Dennet, Daniel C. 1991. *Consciousness Explained.* Boston: Little, Brown.

Dillenberger, John, editor. 1962. *Martin Luther Selections from His Writings.* New York: Anchor Books Doubleday.

Dossey, Larry. 1989. *Recovering the Soul A Scientific and Spiritual Search.* New York: Bantam Books.

Eccles, John C. 1994. *How The Self Controls Its Brain.* Berlin-Heidelberg-New York: Springer-Verlag.

Feynman, Richard. 1965, *The Character of Physical Law..* Cambridge, MA: MIT Press.

Filkin, David. 1997. *Stephen Hawking's Universe: the Cosmos Explained.* New York: BasicBooks HarperCollins Publishers.

Frankl, Viktor E. 1959, 1984 ed. *Man's Search for Meaning.* New York: Washington Square Press.

Frankl, Viktor E. 1969, 1988 ed. *The Will to Meaning Foundations and Applications of Logotherapy.* New York: Penguin Books.

Gazzaniga, Michael S. 1988. *Mind Matters How Mind and Brain Interact to Form Our Conscious Lives.* Boston: Houghton Mifflin Company.

Glynn, Patrick. 1997. *God: the Evidence: The Reconciliation of Faith and Reason in a Postsecular World.* Rocklin, CA: Prima Publishing.

Goldman, Steven L. ed. 1989. *Science, Technology, and Social Progress.* Bethlehem: Lehigh University Press; London and Toronto: Associated University Presses.

Greene, Brian R. 1999. *The Elegant Universe: Superstrings, Hidden Dimensions, and the Quest for the Ultimate Theory.* New York: W. W. Norton & Company.

Grube, G. M. A. 1981. *Plato Five Dialogues Euthyphro, Apology, Crito, Meno Phaedo.* Indianapolis: Hackett Publishing Company.

Guthrie, W. K. 1971. *Socrates.* Cambridge and New York, New York: Cambridge University Press.

Halliday, David and Robert Resnick. 1974. *Fundamentals of Physics.* New York: John Wiley & Sons.

Harth, Erich. 1993. *The Creative Loop: How The Brain Makes A Mind.* Reading, MA: Addison-Wesley.

Hawking, Stephen. 1988, 1996 ed. *A Brief History of Time from the Big Bang to Black Holes.* New York: A Bantam Book.

Hawking, Stephen. 2001. *The Universe in a Nutshell.* New York: Bantam Books.

Hecht, Eugene. 1998. *Physics: Algebra/Trig.* New York: Adelpi University, Brooks/Cole Publishing Company.

Heisenberg, Werner. 1971. *Physics and Beyond: Encounters and Coversations.* New York: Harper & Row, Publishers, Inc.

Hodgson, David. 1991. *The Mind Matters: Consciousness and Choice in a Quantum World.* Oxford: Clarendon Press.

Hogan, James P. 1997. *Mind Matters: Exploring the World of Artificial Intelligence.* New York: The Ballantine Publishing Group.

Hoyle, Fred. 1983. *The Intelligent Universe.* New York: Holt, Rinehart and Winston.

Hook, Sidney, editor. 1958. *Determinism and Freedom in the Age of Modern Science.* A Philosophical Symposium. New York:University Press, Washington Square.

Hume, David. "Of Liberty and Necessity" Part 1. In *An Enquiry Concerning Human Understanding.*

Johnson, R. Colin and Chappell Brown. 1988. *Cognizers: Neural Networks and Machines That Think.* New York: Wiley Science Editions John Wiley & Sons, Inc.

Johnson, Phillip E. 1991. *Darwin on Trial.* Washington, DC: Regenery Gateway.

Kaku, Michio and Jennifer Thompson. 1995. *Beyond Einstein: the Cosmic Quest for the Theory of the Universe.* Anchor Books Doubleday.

Koestler, Arthur. 1967, 1976 ed. *The Ghost in the Machine.* New York: Random House.

Kosslyn, Stephen Michael and Olivier Koenig. 1992. *Wet Mind: The New Cognitive Neuroscience.* New York: The Free Press A Division of Macmillan, Inc.

Krauss, Lawrence. 1995. *The Physics of Star Trek.* New York: Basic Books HarperCollins Publishers, Inc.

Krauss, Lawrence. 1997. *Beyond Star Trek: Physics from Alien Invasions to the End of Time.* New York: HarperCollins Publishers, Inc.

Kurzweil, Ray. 1999. *The Age of Spiritual Machines When Computers Exceed Human Intelligence.* New York: Penguin Books.

Libet, Benjamin, Anthony Freeman, and Keith Sutherland, eds. 1999. *The Volitional Brain Towards a Neuroscience of Free Will.* Thorverton, UK: Imprint Academic.

Luther, Martin. 1957. *The Bondage of the Will.* Translated by J. I. Packer and A. R. Johnston. London: James Clarke and Co. Ltd. Westwood, N.J.: Fleming H. Revell Company.

Margenau, Henry. 1968. *Scientific Indeterminism and Human Freedom: Wimmer Lecture XX.* Fleming H. Revell Company: The Archabbey Press.

Margenau, Henry. 1984. *The Miracle of Existence.* Woodbridge, Connecticut: Ox Bow Press.

Margenau, Henry and Roy Abraham Varghese, editors. 1992. *Cosmos, Bios, Theos.* La Salle, Illinois: Open Court Publishing Company.

May, Rollo. 1969. *Love and Will.* New York: Norton, Dell Publishing.

McCulloch, Warren S. 1965. *Embodiments of Mind.* Cambridge, MA: The M. I. T. Press.

Miller, Keneth R. 1999. *Finding Darwin's God: A Scientist's Search for Common Ground between God and Evolution.* New York: Harper Collins Publishers.

Minsky, Marvin. 1985. *The Society of Mind.* New York: Simon and Schuster.

Mitchell, Stephen A. 1995. *Freud and Beyond: A History of Modern Psychoanalytic Thought.* New York: Basic Books.

Moreland, J. P., editor. 1994. *The Creation Hypothesis – Scientific Evidence for an Intelligent Designer.* Downers Grove, IL: InterVarsity Press.

Murphy, Joseph. 1963. *The Power of Your Subconscious Mind.* New York: Bantam Books.

Nadeau, Robert and Menas Kafatos. 1999. *The Non-Local Universe: The New Physics and Matters of the Mind.* Oxford: Oxford University Press.

Nature 386. April 3, 1997. Larson, Edward J. and Larry Witham. "Scientists Are Still Keeping the Faith."

Palmer, Donald. 1994. *Looking at Philosophy.* Mountain View, CA: Mayfield Publishing Company.

Pennock, Robert, ed. 2001. *Intelligent Design Creation and Its Critics, Philosophical Theological and Scientific Perspectives.* Cambridge, MA: A Bradford Book, The MIT Press.

Penrose, Roger. 1994. *Shadows of the Mind – A Search for the Missing Science of Consciousness.* Oxford, New York, Melbourne: Oxford University Press.

Perkowitz, Sidney. 1996. *Empire of Light, A History of Discovery in Science and Art.* New York: Henry Holt and Company, Inc.

Plato. 1928. Irwin Edman, editor. *The Works of Plato.* New York: The Modern Library Random House.

Restak, Richard, M.D. 1991. *The Brain Has a Mind of Its Own Insights from a Practicing Neurologist.* New York: Harmony Books Crown Publishers, Inc.

Richards, Robert J. 1987. *Darwin and the Emergence of Evolutionary Theories of Mind and Behavior.* Chicago: The University of Chicago Press.

Ryle, Gilbert. 1949, 1984 ed. *The Concept of Mind.* Chicago: University of Chicago Press.

Sagan, Carl. 1973. *The Cosmic Connection An Extraterrestrial Perspective.* New York: Doubleday.

Sagan, Carl. 1977. *The Dragons of Eden.* New York: Random House.

Sagan, Carl and Ann Druyan. 1992. *Shadows of Forgotten Ancestors.* New York: Ballantine Books.

Sagan, Carl. 1996. *The Demon-Haunted World Science as a Candle in the Dark.* New York: Random House.

Sagan, Carl. 1997. *Billions and Billions: Thoughts on Life and Death at the Brink of the Millennium.* New York: Random House, Inc.

Schrödinger, Erwin. 1944, 1992 ed. *What is Life?* Cambridge: Cambridge University Press.

Schroeder, Gerald, L. 1997. *The Science of God: The Convergence of Scientific and Biblical Wisdom.* New York: The Free Press.

Schwartz, Jeffrey M. and Sharon Begley. 2002. *The Mind and The Brain Neuroplasticity and the Power of Mental Force.* New York: Harper-Collins Publishers Inc.

Searle, John R. 1984. *Minds, Brains and Science.* London: British Broadcasting Corporation.

Searle, John R. 1992. *The Rediscovery of the Mind.* Cambridge, MA: MIT Press.

Searle, John R. 1997. *The Mystery of Consciousness.* New York: The New York Review of Books.

Sekuler, Robert and Randolph Blake. 1998. *Star Trek on the Brain: Alien Minds, Human Minds.* New York: W. H. Freeman and Company.

Sigmund, Paul E., translator and editor. 1988. *St. Thomas Aquinas on Politics and Ethics.* New York: Princeton University, W. W. Norton & Company.

Silverstein, Alvin and Virginia. 1986. *World of the Brain.* New York: William Morrow and Co.

Smolin, Lee. 1997. *The Life of the Cosmos.* New York: Oxford University Press.

Spielberg, Nathan and Bryon Anderson. 1995. *Seven Ideas that Shook the Universe.* New York: John Wiley & Sons, Inc.

Sproul, R. C. 1997. *Willing to Believe, The Controversy over Free Will.* Grand Rapids, MI: Baker Books.

Sproul, R. C. 2000. *The Consequences of Ideas Understanding the Concepts that Shaped our World.* Wheaton, IL: Crossway Books, a division of Good News Publishers.

Stapp, Henry P. 1993. *Mind, Matter, and Quantum Mechanics.* New York: Springer-Verlag.

Stenger, Victor. 1995. *The Unconscious Quantum: Metaphysics in Modern Physics and Cosmology.* Amherst, New York: Prometheus Books.

Swinburne, Richard. 1996. *Is There a God?* New York: Oxford University Press Inc.

Thornton, Mark. 1989. *Do We Have Free Will?* New York: St. Martin's Press, Inc.

Tilby, Angela. 1992. *Soul: God, Self, and The New Cosmology.* New York: Doubleday.

Van de Weyer, Robert. 1997. *Socrates in a Nutshell.* London: Hodder & Stoughton.

Vendler, Zeno . 1972. *Res Cogitans An Essay in Rational Psychology.* Ithaca, NY: Cornell University Press.

Von Neumann, John. 1958. *The Computer and the Brain.* New Haven: Yale University Press.

Watson, J. B. 1924, 1970 ed. *Behaviorism.* New York: W. W. Norton & Company.

Weinberg, Steven. 1977, 1993 ed. *The First Three Minutes A Modern View of the Origin of the Universe.* New York: Basic Books.

Weinberg, Steven. 1992. *Dreams of A Final Theory.* New York: Vintage Books, A Division of Random House, Inc.

Wells, Johnathan. 2000. *Icons of Evolution: Science or Myth?: Why Much of What We Teach about Evolution Is Wrong.* Washington, D. C.: Regnery Publishers.

Wolf, Fred Alan. 1996. *The Spiritual Universe: How Quantum Physics Proves the Existence of the Soul.* New York: Simon and Schuster.

Yeffeth, Glenn, ed. 2003. *Taking the Red Pill: Science, Philosophy and Religion in the Matrix.* Dallas, Texas: Benbella Books.

Zohar, Danah. 1990. *The Quantum Self: Human Nature and Consciousness Defined by the New Physics.* New York: William Morrow and Company, Inc.

Zohar, Danah and Dr. Ian Marshall. 2000. *SQ: Connecting with Our Spiritual Intelligence.* New York: Bloomsbury Publishing.

Concordance